Oliver Dehus

Receptor polymorphisms and non-classical bacterial immune stimuli

Oliver Dehus

Receptor polymorphisms and non-classical bacterial immune stimuli

Variants of the human TLR4 and of the bacterial surface molecules LPS and LTA - their specific roles in innate immunity

Südwestdeutscher Verlag für Hochschulschriften

Impressum/Imprint (nur für Deutschland/ only for Germany)
Bibliografische Information der Deutschen Nationalbibliothek: Die Deutsche Nationalbibliothek verzeichnet diese Publikation in der Deutschen Nationalbibliografie; detaillierte bibliografische Daten sind im Internet über http://dnb.d-nb.de abrufbar.

Alle in diesem Buch genannten Marken und Produktnamen unterliegen warenzeichen-, markenoder patentrechtlichem Schutz bzw. sind Warenzeichen oder eingetragene Warenzeichen der jeweiligen Inhaber. Die Wiedergabe von Marken, Produktnamen, Gebrauchsnamen, Handelsnamen, Warenbezeichnungen u.s.w. in diesem Werk berechtigt auch ohne besondere Kennzeichnung nicht zu der Annahme, dass solche Namen im Sinne der Warenzeichen- und Markenschutzgesetzgebung als frei zu betrachten wären und daher von jedermann benutzt werden dürften.

Verlag: Südwestdeutscher Verlag für Hochschulschriften Aktiengesellschaft & Co. KG
Dudweiler Landstr. 99, 66123 Saarbrücken, Deutschland
Telefon +49 681 37 20 271-1, Telefax +49 681 37 20 271-0, Email: info@svh-verlag.de
Zugl.: Constance, University, Dissertation, 2008

Herstellung in Deutschland:
Schaltungsdienst Lange o.H.G., Zehrensdorfer Str. 11, D-12277 Berlin
Books on Demand GmbH, Gutenbergring 53, D-22848 Norderstedt
Reha GmbH, Dudweiler Landstr. 99, D- 66123 Saarbrücken
ISBN: 978-3-8381-0654-0

Imprint (only for USA, GB)
Bibliographic information published by the Deutsche Nationalbibliothek: The Deutsche Nationalbibliothek lists this publication in the Deutsche Nationalbibliografie; detailed bibliographic data are available in the Internet at http://dnb.d-nb.de.

Any brand names and product names mentioned in this book are subject to trademark, brand or patent protection and are trademarks or registered trademarks of their respective holders. The use of brand names, product names, common names, trade names, product descriptions etc. even without
a particular marking in this works is in no way to be construed to mean that such names may be regarded as unrestricted in respect of trademark and brand protection legislation and could thus be used by anyone.

Publisher:
Südwestdeutscher Verlag für Hochschulschriften Aktiengesellschaft & Co. KG
Dudweiler Landstr. 99, 66123 Saarbrücken, Germany
Phone +49 681 37 20 271-1, Fax +49 681 37 20 271-0, Email: info@svh-verlag.de

Copyright © 2008 Südwestdeutscher Verlag für Hochschulschriften Aktiengesellschaft & Co. KG and licensors
All rights reserved. Saarbrücken 2008

Produced in USA and UK by:
Lightning Source Inc., 1246 Heil Quaker Blvd., La Vergne, TN 37086, USA
Lightning Source UK Ltd., Chapter House, Pitfield, Kiln Farm, Milton Keynes, MK11 3LW, GB
BookSurge, 7290 B. Investment Drive, North Charleston, SC 29418, USA
ISBN: 978-3-8381-0654-0

Table of content

1 Introduction 3
   1.1   Infection and innate immunity 3
   1.2   Cytokines in inflammatory responses 4
   1.3   Immune recognition by toll-like receptors 5
   1.4   The polymorphism Asp(299)Gly of TLR4 6
   1.5   Major bacterial immune stimuli 7

2 Aims of the study 13

3 IL-10 release requires stronger toll-like receptor 4-triggering than TNF-α possible explanation for the selective effects of heterozygous TLR4 polymorphism Asp(299)Gly on IL-10 release 15
   3.1   Abstract 15
   3.2   Introduction 15
   3.3   Material and Methods 16
   3.4   Results 18
   3.5   Discussion 22

4 LPS-inducible anti-inflammatory responses are not diminished in Crohn's disease patients with heterozygous TLR4 Asp(299)Gly polymorphism 25
   4.1   Abstract 25
   4.2   Introduction 26
   4.3   Material and Methods 26
   4.4   Results 27
   4.5   Discussion 28
   4.6   Acknowledgements 29

5 Endotoxin evaluation of eleven lipopolysaccharide by whole blood assay does not always correlate with Limulus Amebocyte Lysate assay 31
   5.1   Summary 31

5.2	Introduction	32
5.3	Material and Methods	33
5.4	Results	35
5.5	Discussion	41
5.6	Acknowledgements	44

6 Growth temperature induces switching of structural variants of *Listeria monocytogenes* lipoteichoic acid 45

6.1	Summary	45
6.2	Introduction	46
6.3	Materials and Methods	47
6.4	Results	51
6.5	Discussion	58
6.6	Acknowledgements	61

7 Summarizing Discussion 63

8 Summary 69

9 Abbreviations 71

10 References 73

11 List of publications 89

12 Acknowledgements 90

1 Introduction

1.1 Infection and innate immunity

Infectious diseases are globally the main cause of death. They are caused by microorganisms that colonize and intrude into the host where they spread and replicate, accompanied with characteristic symptoms of inflammation. Our innate immune system, described over a century ago, is derived from the phylogenetically oldest defence mechanisms and has been conserved in all multicellular organisms, constituting the organism's first line of defence against invading pathogens [1, 2]. The microbe-host interaction is a paradigm for co-evolutional events that appear as a competition between infectious agents which are continuously optimising their invasive strategies on the one side and the defence mechanisms of the host on the other side. The immune system needs to balance these mechanisms to efficiently eradicate the pathogens but at the same time avoiding deleterious effects for the body. The white blood cells, which are the major players, derive from myeloid precursors and include competent phagocytes: monocytes/macrophages, dendritic cells (DC) and neutrophilic granulocytes (PMN). Initial detection of pathogens involves the pattern recognition receptors (PRR) of immune cells and also other cell types. These immune receptors are expressed on the cell surface, in organelles and in the cytosol sensing the presence of pathogens by their microbe associated molecular patterns (MAMPs, formerly PAMPs) [3]. MAMPS are indispensable molecules whose functions are sensitive to variations and therefore possess highly conserved structures [4, 5]. Upon stimulation by receptor-ligand binding, inflammatory effector substances like cytokines, lipid mediators and nitric oxide are released from monocytes/macrophages, stimulating the activity of several populations of target cells [6, 7]. Amongst those, endothelial cells and immune cells respond with the expression of surface adhesion molecules, phagocytes are activated for lysosomal clearance and eliminate pathogens by respiratory burst, DC and monocytes/macrophages express co-stimulatory molecules and differentiate to antigen-presenting cells stimulating the specific immune system involving B- and T- lymphocytes [8]. Apart from the cell-mediated response, an innate humoral defence mechanism involving a variety of serum proteins is known as the complement pathway, which can be activated by three different manners: the classical pathway initiated by IgM and IgG antibodies bound to their peptide antigens and quite similar, the lektin pathway initiated by leucin-rich repeats-carrying recognition molecules like mannose binding lektin bound to certain sugar moieties. The alternative pathway gets activated by spontaneous hydrolysis of effector molecules in the presence of a variety of microbial cell wall components. Activation of complement initiates zymogenic cascades, releasing inflammatory mediators as well as fragments for the opsonisation or direct inactivation of pathogens and toxins [9].

Regarding host-pathogen interactions and inflammatory responses, broad inter-individual variations exist and have to be clarified for adequate treatments. However, the courses of diseases are too complex to let us understand completely how the same pathogen causes subclinical, mild, severe or lethal infections. The status of the immune system, which critically depends on the physical condition, determines the outcome of an infection [10]. Deviations from an optimal immune response, which would prevent the invasion and spread of pathogens, might result in an overwhelming inflammatory reaction damaging the body or a diffident, insufficient defence. Possible explanations for deviations in immune responses are e.g. genetic polymorphisms or regulatory dysbalances occurring for instance during immune suppression. Several individual divergences were already shown to be due to genetic or epi-genetic predisposition like gender, ethnical origin and age [11-14]. Furthermore, the pathogens themselves represent a tremendous source of variability, including structural and strategical variations challenging the immune system. In order to develop effective strategies of prevention and therapy, a much more profound understanding of the molecular mechanisms and its variations involved in immune recognition and signalling on the host's side and virulence on the microbes' side are necessary with intelligent though reductionistic model systems to be the key.

1.2 Cytokines in inflammatory responses

Ineffective recognition of pathogens or inappropriate immune responses may lead to uncontrolled microbial growth or overwhelming systemic inflammatory responses followed by tissue damage, vascular collapse and multiorgan failure, as occurring in severe sepsis and septic shock [15]. However, potent endotoxins alone like lipopolysaccharide (LPS) of Gram-negative bacteria are capable of triggering adverse clinical responses, including procoagulant response and septic shock [16]. Taveira da Silva *et al.* could demonstrate, that self-administration of Salmonella endotoxin mimics many of the clinical features of septic shock [17]. The most common microbes isolated from patients with severe Gram-negative sepsis are *Escherichia coli*, *Klebsiella* species and *Pseudomonas aeruginosa* [18], whereas *Listeria monocytogenes* is a prominent cause of Gram-positive sepsis [88].

In sepsis, the prevailing cytokines that are involved in a systemic response are tumour necroses factor (TNF), interleukin (IL)-1β and IL-6 which cause hypotension and organ failure associated with lethal septic shock [19, 20]. Monocytes/macrophages constitute the principal source of proinflammatory cytokines elicited by endotoxins [21].

As a counter-player, IL-10 represents one of the most important immune-regulating cytokines and is mainly expressed by monocytes/macrophages, but also in lymphocytes, mast cells and other cell types. It confers its mainly immunosuppressive effects on the immune cells of both the innate and the adaptive immune system. In macrophages, IL-10 release follows TNF production, and down-regulates the proinflammatory reactions [22]. Above all, the inhibition of TNF, IL-1β and IL-6 is crucial, because these cytokines have synergistic activities on inflammatory processes and amplify these responses by inducing secondary mediators such as chemokines, eicosanoids and platelet activating factor. The IL-10 induced inhibition of inflammation is mediated by modulation of transcription and reduction of the stability of mRNA, characterized by AU-rich elements in the 3'-untranslated regions [23, 24]. Furthermore, IL-10 leads to inhibition of NFκB, which plays a key role as transcription factor for many inflammatory genes, via suppressing both inhibitor κ B kinase and DNA binding of NFκB [25]. IL-10 also enhances the production of the antagonists of some proinflammatory effectors, e.g. of soluble p55 and p75 TNF receptor [26, 27] as well as IL-1 receptor antagonist [28]. The anti-inflammatory potential of IL-10 has been demonstrated by preventing experimental endotoxaemia [29, 30] and suppressing experimental intestinal inflammation in the mouse [31]. The severe consequences of a misbalance of circulating proinflammatory and anti-inflammatory cytokines become evident in trauma and sepsis patients where the IL-10 to TNF ratio is high [32-34]. The other extreme, i.e. a low IL-10 to TNF ratio, is associated with an imbalance in favour of proinflammatory cytokines, as observed in case of autoimmune diseases, e.g. of systemic lupus erythematodus [35] or inflammatory bowel disease which is characterized by chronic mucosal inflammation, a possible consequence of a dysbalance of proinflammatory and regulatory cytokines [36, 37]. In some cases, the benefit of IL-10 therapies is discussed [38, 39].

1.3   Immune recognition by toll-like receptors

With at least 11 different members identified, the toll-like receptors (TLRs) form the greatest family of PRRs and are of major significance for the detection of MAMPs in mammals [3, 40, 41]. In 1988, the Toll protein was first described in Drosophila, where it initiates immune responses against fungal infections [42, 43]. Subsequently, a set of mammalian proteins containing an extracellular c-terminal leucin-rich repeat and an intracellular N-terminal toll/interleukin-1 receptor (TIR) domain were found to be structural related to Drosophila toll and called TLRs [44]. The TLRs detect a variety of different PAMPs, including e.g. tri-acyl lipopeptides from bacteria and mycobacteria (TLR1) [45, 46], lipoproteins and lipo-teichoic acid (LTA) from Gram-positive bacteria (TLR2) [47], double-stranded viral RNA (TLR3) [48], LPS from Gram-negative bacteria (TLR4) [49, 50], flagellin (TLR5) [51], di-acyl lipopeptides from mycoplasma (TLR6) [52], GU rich single strand RNA (TLR7 and TLR8) [53] and bacterial DNA (TLR9) [54]. TLR10 [55] and recently TLR11 [41] have also been identified, but their ligands are unknown so far. The first human TLR described was TLR4 [56] and the *tlr4* gene was identified in 1998 [57]. A defect of TLR4 mediated signalling in C3H/HeJ mice due to a point mutation was found to result in an incapability of responding to LPS [49, 50]. Human TLR4, located on chromosome 9, is expressed by monocytes/macrophages, DC, PMN, mast cells and at organ-specific levels by epithelial cells [58-61]. During the initiation of an immune response which is in first place initiated by the PRRs, the pattern of cytokines released by immune cells is crucial for a successful host defence and varies depending not only on the pathogen involved, but also on the individual host. In the latter context, mutations in form of single nucleotide polymorphisms (SNP) entailing an altered immune recognition are currently a major matter of research since they might account for inter-individual susceptibilities towards certain diseases and for differences in immune reactions; one prominent example are the polymorphisms of TLR4.

1.4 The polymorphism Asp(299)Gly of TLR4

Genetic polymorphisms are allelic variants within a population occurring by definition at a frequency of over 1%. Their most common appearances are SNPs, which can either be silent or result in a functional aberration. If located in a promotor region, SNPs may affect gene expression or in case of an amino acid exchange alter the protein structure. According to the hypothesis of mutations emerging in genomic regions of strong selective pressure, genes involved in immunity and particular in immune recognition exhibit a relatively high number of SNPs [62]. In this respect, the question why significant inter-individual differences in susceptibility to infection and its severe outcomes exist, is freshly discussed [31, 63].

Since Gram-negative infections are still of outstanding clinical relevance, many efforts have been undertaken to precisely elucidate the role of the LPS recognition receptor TLR4 and its polymorphisms. Arbour et al. screened the coding region of the *tlr4* gene and detected two co-segregating missense mutations that affect the extracellular domain [64], which is considered the most conserved one [65].

In one case, an A to G transition at position +896 downstream of the translation start site results in the replacement of an aspartic acid residue (Asp) by glycine (Gly) at amino acid 299. The second polymorphism was found to be in complete linkage disequilibrium with *tlr4*/A(896)G and was a C to T transition at position +1196, resulting in an exchange of threonine (Thr) by isoleucine (Ile) at amino acid 399. In transfection experiments, the Asp(299)Gly but not the Thr(399)Ile mutation was found to interrupt LPS-induced TLR4 signalling [64]. In numerous studies with patients, some associations of the Asp(299)Gly polymorphism with several diseases have already been reported, suggesting that the mutation might in deed alter inflammatory responses [62]. Moreover, variations of the sensitively balanced inflammatory actions during immune responses are believed to be linked with an increased susceptibility towards the development of chronic diseases like e.g. autoimmune disorders, asthma or inflammatory bowel disease; in the latter case, patients suffering from Crohn's disease or ulcerative colitis have been found to carry the heterozygous TLR4 Asp(299)Gly polymorphism at a increased frequency [66, 67]. However, no functional studies concerning the effect of this mutation for LPS-induced immune responses have been performed so far.

1.5   Major bacterial immune stimuli

1.5.1   Bacterial pathogens

The majority of infectious agents relevant for humans are found in the domain of the prokaryotes. Despite the introduction of antibiotics, infections with extracellular or intracellular replicating bacteria are steadily increasing [68] and mortality of bacteraemia remains high with 25-40% [69-71]. Especially in non-industrialized countries, infectious diseases like gut infections are still the main cause of mortality and morbidity. In general, symbiotic bacteria constitute the individual human body flora with 500 to 1000 species, performing indispensable metabolic tasks and avoiding the establishment of pathogenic micro-organisms. However, amongst these commensals opportunistic pathogens exist that may cause an infection when they get the chance to become invasive, like in immunocompromised individuals [72]. Invasion involves a complex, and in many cases poorly understood activation of virulence factors; some of those are also responsible for the adaptation to a physically and chemically different environment and for immune evasion. One typical regulatory mechanism is the two-component system which senses extracellular changes like temperature or osmolarity and induces the display of defined genetic programs organised in regulons. Variability and horizontal gene transfer fosters the efficiency and spread of virulence genes [73]. Furthermore, both pathogenic bacteria and even non-pathogenic symbiotics possess the ability to express molecules that cause after entering the blood stream, inflammation and provoke symptoms of sepsis. Such immunogenic substances can either be secreted (e.g. Listeriolysin O, Staphylococcal enterotoxin B) or released after cell death, or cell-renewal and -division. Thus, some of the most immunogenic compounds recognized by TLRs are expressed in the cell wall (e.g. lipoproteins, LPS, LTA). The cell wall of the prokaryotes is a flexible but robust building which withstands the turgor and shields the organism from many substances with antibiotic activities. For that reason, its turnover is carefully regulated to ensure growth and cell division without damage. These circumstances as well as the fact that the cell wall contains the bacteria's outermost components contacting and interacting with the host, have made the bacterial cell wall an intensively studied subject–probably harbouring the key to at least transiently overcome the massive health problems due to increasing antibiotic resistances. Within the domain of prokaryotes, a common classification is done by the feasibility of Gram staining, thus discriminating Gram-negative from Gram- positive bacteria due to differences in the cell wall. While Gram-negatives possess two phospho-lipid membranes with a thin layer of peptidoglycan in between, Gram-positives have only one phospho- lipid membrane surrounded by a thick layer of peptidoglycan. Besides the immunogenic components of the peptidoglycan and lipoproteins, Gram- negative bacteria express the highly potent LPS while Gram-positives express LTA which provokes a moderate inflammatory response [74, 75]. Typical for MAMPs, both LPS and LTA show a highly conserved structure building up a repetitive hydrophilic chain participating in forming the cell surface connected to a lipid moiety which is embedded in the membrane [76].

1.5.2 Lipopolysaccharide

The basic structure of classical, "smooth" LPS as examined mostly for enterobacteriaceae consists of a repetitive polysaccharide chain (O-antigen) with a high variability determining the serological specificity, the core oligosaccharide and a lipid moiety, also named lipid A (Fig. 2 A). This β-1,6-linked D-glucosamine disaccharide bearing two phosphate groups in position 1' and 4', substituted with six fatty acids 12 to 14 carbons in length, is alone sufficient for the activation of TLR4 mediated signalling and full toxic activity *in vivo* and *in vitro* [77-80]. LPS induces the expression of a cytokine pattern similar to stimulation with whole bacteria and also activates the complement system [80, 81]. Deviations from the architecture of the prototypical LPS have been identified only recently, like the phosphorylation pattern of the disaccharide or the number and nature of the acyl chains. Such deviations occur only in a limited range, but bear significant consequences for the immune recognition, concerning the recognition by specific PRR and the induction of cytokine patterns [77, 82, 83, 150]. Such exceptions from the rule are the penta-acyl lipid A with partially unsaturated carboxylic acid residues from *Rhodobacter sphaeroides* (LPS-receptor TLR4 antagonist, [84]), the O-methylized monophosphorylated lipid A from *Leptospira interrogans* (TLR2 antagonist, [85] or the monophosphorylated penta-acyl LPS from *Porphyromonas gingivalis* (signalling via lipoprotein receptor TLR2; [86]). For the opportunistic pathogen *Pseudomonas aeruginosa* it was shown recently that the acylation of the lipid A can differ between isolates from the environment or from a source of cystic fibrosis, associated with different immune stimulatory potencies [87]. In all cases of non-classical lipid A structures, the induction of cytokine release from blood leukocytes is less potent. Taken together, the system of PRR sensing LPS seems to display a very specific receptor-ligand interaction which is sensitive towards even small sterical modifications. Still, many questions concerning the association between non-classical LPS architectures and immune recognition have to be solved in order to understand the species-specific infection strategies and inflammatory responses.

1.5.3   Lipo-teichoic acid

Until recently, the majority of infectious diseases were referred to the Gram-negative bacteria and research had been focusing on them and on their highly pyrogenic LPS and its lipid A moiety. However, today Gram-positive infections are increasing, first of all in immunocompromised individuals [88] and consequently those cell wall components of Gram-positives that are indispensably involved in bacterial life and pathogenity are being examined more closely. LTA is a molecule apparently combining those two aspects: Its heterogenous functions comprise colonisation, cell division and virulence [89-91]- involving the regulation of autolytical activity, homeostasis of physiochemical surface properties [92], cation homeostasis [93] and resistance to antimicrobial cationic molecules [94]. In the opportunistic intracellular pathogen *Listeria monocytogenes*, LTA is reported to be the scaffold for non-covalently bound proteins like internalin B (InlB) which alone is able to confer invasiveness into host cells [95, 96]. At the same time, LTA is a unique stimulus of cytokines, inducing a strong chemokine expression but almost no IL-12 or IFNγ [74, 97]. Furthermore, LTA activates the L-ficolin dependent C4 turnover of complement [98]. Not only the immune stimulatory capacity, but also the amphiphilic structure of LTA resembles its Gram-negative "counterpart" LPS.

The well characterized LTA from *Staphylococcus aureus* is made up of a polyglycerophosphate backbone with in average 48 repeating units, substituted with D-alanine (70%) and α-D-N-acetylglucosamine (15%). This backbone, protruding the cell wall, is connected via a gentiobiose (α 1-6 glucose β) to a membrane-anchored diacylglycerol, containing 50% methylated fatty acid residues with an average length of 14 carbons [99]. Structural deviations concerning the backbone length, its substituents, the disaccharide and the length of the fatty acid residues are already known for the LTA from *S. pneumoniae* ([100], *B. subilits* [97] or Lactobacilli species [101], but are not associated with significant differences in immune stimulation [102]. By the use of synthetic LTA derivatives it could be shown that the diacylglycerol alone displays weak biological activity, while a complete cytokine release compared to native LTA was induced when six glycerophosphate units substituted with four D-alanine and one D-N-acetylglucosamine were connected to the diacylglycerol via a gentiobiose [74, 103]. Since until now no natural LTA mutants are known, the importance of LTA for the bacteria is obvious. Its functional variety and its immune stimulatory potency make it a promising molecule for investigating pathogen-host interactions and adaptation strategies in order to develop of bactericidal treatments, possibly interfering with LTA synthesis

1.5.4 *Listeria monocytogenes* as an intracellular pathogen

Intracellular bacteria independent whether they are facultative or obligatory pathogens, are in contrast to the obligate extracellular ones not limited to the epithelium of the host but become invasive. The intracellular immune recognition and responses are poorly elucidated until now. However, the cytosolic PRRs NOD1 and NOD2 have been reported recently to sense the presence of muropeptides, fragments from the cell wall peptidoglycan [104-106]. In some cases, like *Listeria monocytogenes* and *Shigella spp.*, the attacks of the humoral immune response are avoided by direct cell to cell spread, making a protective host defence depending on the T-cell responses necessary. They become internalized into the host cells via zipper- or trigger mechanisms and thus are localised in endosomes or phagosomes. To overcome the bactericidal medium of those vacuoles, different strategies have evolved to either escape into the cytosol (e.g. *L. monocytogenes, Shigella spp.*), become resistant (e.g. *Francisella tularensis, Salmonella typhimurium*) or render the vacuoles harmless (e.g. *Legionella pneumophila, Mycobacteria spp.*). After having crossed the human epithelial barrier they may infiltrate lymphatic tissue and from there be transported to the spleen and the liver. Via the lymph- and the bloodstream virtually all organs like kidneys, lung, heart or brain can become infected. Whereas some pathogens are specialized for colonizing certain organs (e.g. *Shigella flexneri* in the colon), others lead to general systemic infections. A prominent example that has become a model organism for studying pathogen-host interactions is the opportunistic intracellular Gram-positive rod *L. monocytogenes* [107]. Being the only known human pathogenic strain of the *Listeria* genus, *L. monocytogenes* is detected on 15% of the foodstuff including vegetables, meat and seafood, making it a transient inhabitant of the human and animal gastrointestinal tract with estimated five to nine exposures per person and year, therewith providing the basis for gut invasive listeriosis [88, 108-110]. Compared to other food born diseases, systemic listeriosis is relatively rare: 1< 100.000 per year in Germany but the lethality of 25-30% is much higher than for other gut infections like salmonellosis. For hosts with a non-competent immune system, the risk is high: AIDS patients are 300 times more susceptible than the average population and unborn children have almost no change for survival if therapy is delayed [111, 112].

The mechanisms of infection and intracellular growth have been investigated in several cellines, including epithelial cells, macrophages and hepatocytes, showing a relatively similar replication cycle of *L. monocytogenes*. The bacteria get internalized by macrophages via phagocytosis or by non-phagocytes via induced phagocytosis involving first of all the internalins (Inl)A and InlB. Within minutes, the phagosomal membrane gets lysed involving the cytolysin Listeriolysin O (LLO) and the phaphatidylinositol-specific phospholipase C (PlcA). In the host cytosol, the Listeria replicate with an average generation time of 40 minutes [88]. Motility is provided by the membrane bound protein ActA which recruits the host protein VASP and katalyses the polarized polymerisation of monomeric G-actin, resulting in the protrution of the bacteria through the host cell. Reaching the plasma-membrane, pseudopodes, also called listeriopodes, trigger the internalization into the vicinal cell. Escaping the vacuole surrounded by a double-membrane, involving additionally the phospholipase C (PlcB), initiates a new cycle. By this cell-to-cell spread, the humoral immune response is avoided. If the innate immune system is not capable of containing invasive Listeria, clearance of infection is then dependent on a T-cell mediated resistance, one explanation for the increased susceptibility towards systemic infections in the case of immunocompromised individuals, pregnant women and their foetuses or newborns. Most of the known genes involved in virulence are regulated by the transcription factor PrfA whose transcriptional and translational expression depends on both physical and chemical factors of the surrounding [88]. The genes that are directly or indirectly affected by PrfA encode for a variety of proteins which mediate into virulence (e.g. host cell entry, phagosomal escape, actin-based motility, hexose-phosphate transport, ABC transport, cell wall modification, secretion [88, 113, 114]. According to what is known about the virulence of Listeria, their immune evasion strategy so far involve the lysosomal escape and the cell to cell spread avoiding humoral defence mechanisms. It is still unclear, what role PRRs play for the sensing of intra- and extracellular Listeria and whether structural modifications of the bacterial cell wall during infection bear further benefits of immune evasion. The capability of sensing and reacting towards changes in the environment is the basis of the bivalent nature of *L. monocytogenes* occurring as an extracellular harmless saprophyte or as an intracellular pathogen. Understanding these regulatory mechanisms and the associated consequences for immune recognition build up a basis for efficient listericidal therapies.

2 Aims of the study

The human innate immune system, faced with severe infectious diseases, is a paradigm for co-evolutional events of a competition between pathogens that are continuously optimising their invasive strategies on the one side and the defence mechanisms that have to balance their powerful force to eradicate infections but at the same time avoiding overwhelming inflammations on the other side. A major characteristicum of the innate immune system is the expression of PRRs, which bind to indispensable microbial molecules whose functions are sensitive to variations and therefore have highly conserved structures. Nevertheless, exceptions from the rules or variations within the natural limits are crucial role for the inter-individual outcome of a certain pathogen/host interaction which may range from subclinical to lethal. For the development of therapeutic or even preventive therapies it is important to understand the deviations in immune recognition which might occur on side of the host cell and on side of the pathogen.

The first part comprises a functional study of the polymorphism Asp(299)Gly of the human pattern recognition receptor TLR4 regarding LPS binding and the induction of the proinflammatory cytokine TNF and the anti-inflammatory cytokine IL-10. Furthermore, a patients study was performed to functionally associate the TLR4 polymorphism with Crohn's disease by examining the LPS-induced IL-10 release in polymorphic versus wild type patients. This section is published or submitted under the titles:

- ▶ IL-10 release requires stronger toll-like receptor 4-triggering than TNF- a possible explanation for the selective effects of heterozygous TLR4 polymorphism Asp(299)Gly on IL-10 release.
- ▶ LPS-inducible anti-inflammatory responses are not diminished in Crohn's disease patients with heterozygous TLR4 Asp(299)Gly polymorphism.

In the second part, non-classical bacterial cell wall molecules with focus on LPS from the opportunistic bacterium *Pseudomonas aeruginosa*, regarding its immune stimulatory potency, and on differentially regulated LTA expression from the extracellular and intracellular grown *Listeria monocytogenes*, are examined. These studies have been published or submitted under the following titles:

- Endotoxin evaluation of eleven lipopolysaccharide by whole blood assay does not always correlate with Limulus Amebocyte Lysate assay.

- Growth temperature induces switching of structural variants of *Listeria monocytogenes* lipoteichoic acid.

IL-10 release requires stronger toll-like receptor 4-triggering than TNF-α possible explanation for the selective effects of heterozygous TLR4 polymorphism Asp(299)Gly on IL-10 release

Oliver Dehus, Sebastian Bunk, Sonja von Aulock, and Corinna Hermann

Biochemical Pharmacology, University of Konstanz, Germany

Immunobiology

3.1 Abstract

The toll-like receptor 4 Asp(299)Gly polymorphism results in an inactive receptor. Heterozygosis is associated with reduced LPS-inducible IL-10 protein and IL-10 mRNA from blood leukocytes and isolated monocytes, while numerous other mediators are not affected. We could exclude that this effect is due to differences in the kinetics of IL-10 release, in the expression of total surface TLR4 or in LPS-binding to monocytes between subjects heterozygous for the Asp(299)Gly polymorphism or homozygous carriers of the wild-type allele. Furthermore, we could show that IL-10 induction in general requires stronger LPS-triggering than TNF and is more sensitive to LPS inhibitors. The lower number of responsive, wildtype TLR4 receptors on monocytes of heterozygotes may explain why only IL-10 release is affected.

3.2 Introduction

The Asp(299)Gly polymorphism of toll-like receptor 4 (TLR4), which mostly co-segregates with the Thr399Ile mutation in Europeans, was found to interrupt lipopolysaccharide (LPS)-induced TLR4 signalling in transfected THP-1 cells and to be associated with reduced responsiveness to inhaled LPS in humans [64]. While two studies about the role of the TLR4 Asp(299)Gly polymorphism in human systemic and peri-operative endotoxemia demonstrated similar responses of subjects with wild-type or heterozygous polymorphic genotype [115, 116], numerous associations with inflammatory or infectious diseases, especially inflammatory bowel disease and Gram-negative infections have been reported [117]. However, in these studies cell-based assays proving that the polymorphism carriers' ability to respond to immune stimuli is altered are mostly lacking. Erridge *et al.* stimulated isolated monocytes with LPS from different Gram-negative bacteria and observed no deficits of the cells from heterozygous TLR4 polymorphism carriers in releasing IL-1β [118]. In a study reported by our group, analysis of cytokine responses of blood leukocytes of 160 healthy volunteers genotyped for the Asp(299)Gly polymorphism in an ex vivo whole blood test did not result in differences in LPS-inducible release of inflammatory mediators like TNF, IL-6, IL-1β, IFNγ, G-CSF, eicosanoids or serum cytokines, except for the release of the anti-inflammatory cytokine IL-10, which was significantly reduced in the group of subjects with heterozygous TLR4 alleles [119].

IL-10 is an important anti-inflammatory cytokine mainly produced by human monocytes, and IL-10 dysfunction can result in excessive inflammation [120]. So far, there is no explanation how this selective effect of the TLR4 Asp(299)Gly polymorphism on IL-10 release is mediated. Here we show that IL-10 production is already reduced on the mRNA level, but the reduced release of IL-10 protein is not due to delayed kinetics. Furthermore, we provide evidence that IL-10 release requires stronger triggering of TLR4 than TNF release, and therefore the lower number of responsive TLR4 receptors on monocytes of heterozygous carriers of the Asp(299)Gly polymorphism may explain why only IL-10 release is affected.

3.3 Material and Methods

3.3.1 Volunteer population and TLR4 genotyping

The TLR4 Asp(299)Gly polymorphism was determined in a population of 558 volunteers recruited at the University of Konstanz, Germany, in the years 2000-2004. DNA was prepared from EDTA anticoagulated blood (Sarstedt) by the QIAamp DNA Blood Mini Kit (Qiagen). Determination of the A(896)G TLR4 SNP was performed by real-time PCR and melting point analysis according to Heesen *et al.* [121]. The heterozygous TLR4 polymorphism occurred with a frequency of 7.2%. Subgroups of subjects with wild-type genotype and heterozygous TLR4 polymorphism were recruited from the 558 volunteers for the investigations described below.

3.3.2 Human whole blood and monocyte incubation

Differential blood cell counts were measured routinely with a Pentra60 to rule out acute infections (ABX Technologies). Incubations of whole blood and ELISA measurements were carried out as described [119]. Monocytes were isolated by MACS-negative selection (Miltenyi Biotec). Stimulations were performed with LPS from *Salmonella abortus equi* (*S.a.e.*), or where indicated with LPS from *Klebsiella pneumoniae* (*Kl.pn.*) (both from Sigma). In some experiments Limulus anti-LPS factor (LALF, a generous gift from F. Jordan, Charles River/Endosafe) was added. RNA from heparinized blood (Sarstedt) was isolated with the QIAamp RNA Blood Mini Kit (Qiagen) and used for reverse transcription. All experiments and measurements were carried out blindly with regard to the donors' genotypes.

3.3.3 Quantitative Real-time PCR

cDNA was quantified by Real-time PCR on a LightCycler system (Roche) with LightCycler FastStart DNA Master SYBR Green (Roche) using specific primers from Thermo Hybaid: TNF forward: 5´-GAGTGACAAGCCTGTAGCCCATGTTGTAGCA-3´, reverse: 5´-GCAATGATCCCAAAGTAGACCTGCCCAGACT-3´; GAPDH forward: 5´-GAAGGTGAAGGTCGGAGTC-3´, reverse: 5´-GAAGATGGTGATGGGATTTC-3´; IL-10 forward: 5´-CAAGTTGTCCAGCTGATCCTTCAT-3´, reverse: 5´-GGCAACCTGCCTAACATG-3´; Cyclophilin forward: 5´-CTCCTTTGAGCTGTTTGCAG-3´, reverse: 5´-GATGGCAAGCATGTGGTG-3´.

3.3.4 FACS analysis

For FACS analysis a FACS Calibur flow cytometer (Becton Dickinson) with Cell Quest software (Becton Dickinson) was used. For assessment of the monocytes' LPS-binding capacity, EDTA blood was stained with in-house produced fluorescein-5 (6)-carboxamido caproic acid N-succinimidyl ester (FCHSE)-LPS and anti-CD14 (BD Biosciences). FCHSE was used as background control. For investigation of TLR4 surface expression, 5×10^5 peripheral blood mononuclear cells (PBMC) were prepared with CPTTM Cell Preparation Tubes (BD Biosciences) and incubated with an anti-TLR4 antibody (a kind gift from Dr. Alexander Dalpke, University of Marburg, Germany). An anti-mouse IgG-phycoerythrin (PE, DAKO) was used as label. Measurement of IgG-PE alone served as background control. Monocytes were gated according to their forward and sideward scattering properties.

3.3.5 Statistics

Statistical analysis was performed using the GraphPad Prism 4.0 program (GraphPad Software, San Diego, USA). Data are given as mean ± SEM. Significance of differences was assessed by t-test in case of two groups only or by one-way ANOVA followed by Bonferroni's post-test. IC_{50} values were determined according to a sigmoidal curve fit.

3.4 Results

The aim of this study was to investigate the selectivity of the effect of the TLR4 Asp(299)Gly polymorphism on IL-10 release. Therefore, we had to compose a newly genotyped study group (n=17 wild-type (+/+); n=10 heterozygous polymorphics (+/-)) and to reconfirm the previously observed effect of the TLR4 Asp(299)Gly polymorphism on LPS-inducible IL-10 release. Again, like in the previous study [119], stimulation of the heterozygous polymorphics' whole blood with LPS (S.a.e. 1 µg/ml) resulted in diminished IL-10 release (+\+: 0.81±0.08 ng/ml vs. +\-: 0.52±0.06 ng/ml, p=0.026), while the release of TNF was not affected (+\+: 2.96±0.38 ng/ml vs. +\-: 3.26±0.44 ng/ml, p>0.05). The same effect was also observed using purified monocytes stimulated with increasing concentrations of LPS (figure 1).

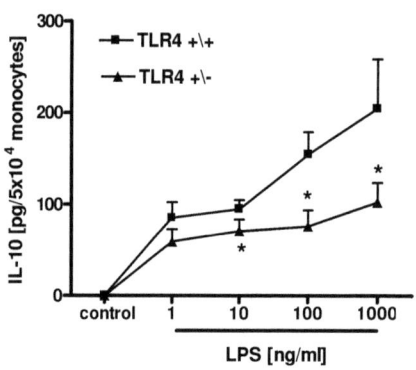

Figure 1 Heterozygous Asp(299)Gly TLR4 polymorphism is associated with reduced IL-10 but not TNF release from isolated human monocytes 5×10^4 monocytes/well were incubated in the presence of LPS at the concentrations indicated for 20h. TNF and IL-10 were determined in the cell-free supernatants by ELISA. TLR4 +\+ indicates the wild-type (n=8) and TLR4 +\- the heterozygous polymorphic genotype (n=7). Data are means ± SEM. *p < 0.05 indicates significance versus the wild-type.

Analysis of mRNA expression by real-time PCR confirmed that lower LPS-inducible IL-10 release occurs already at the IL-10 mRNA level, while TNF mRNA levels were not influenced (figure 2). Since only the IL-10 release was affected by the TLR4 polymorphism, we investigated whether the reduced IL-10 levels were due to a shift in the kinetics of IL-10 release of subjects with TLR4 polymorphisms. For this purpose, we followed the release of LPS-induced IL-10 in whole blood incubations over a period of 28h. We measured IL-10 by ELISA after 5h, 10h, 15h, 20h, 25h and 30h of stimulation, but no shift in the IL-10 release curve was apparent (figure 3).

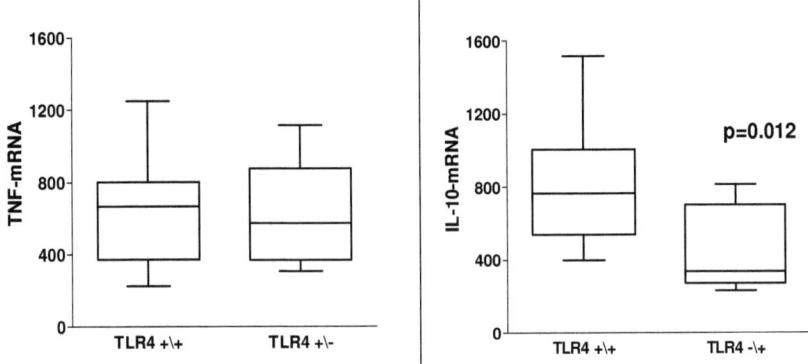

Figure 2 Heterozygous Asp(299)Gly TLR4 polymorphism is associated with reduced IL-10 but not TNF mRNA Five ml of 20% human whole blood were incubated in the presence of 1 µg/ml LPS from S.a.e. for 6h. RNA was prepared, reversely transcribed and cDNA was analyzed by real-time PCR. TNF and IL-10 data were normalized to cyclophilin cDNA. Data are presented in a box and whiskers blot as x-fold induction of mRNA. TLR4 +\+ indicates the wild-type (n=12) and TLR4 +\- the heterozygous polymorphic genotype (n=8).

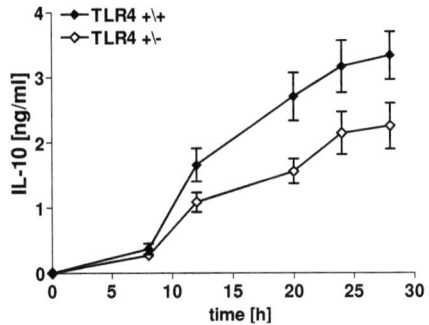

Figure 3 The kinetic of IL-10 release is not affected by the heterozygous Asp(299)Gly TLR4 polymorphism One ml of 20% human whole blood was incubated in the presence of 1 µg/ml LPS from S.a.e. for the time intervalls indicated. IL-10 was determined in the cell-free supernatants by ELISA. Data are means ± SEM. TLR4 +\+ indicates the wild-type (n=14) and TLR4 +\- the heterozygous polymorphic genotype (n=8).

Next we assumed that the difference in the density of total TLR4 surface expression might be responsible for the observed effects. When we compared the total TLR4 surface expression of monocytes from six wild-type and nine heterozygous TLR4 polymorphic donors by FACS analysis, we detected a higher density of TLR4 on monocytes from heterozygous polymorphic donors (median of relative fluorescence: (+\+): 11.92±1.00 vs. (+\-): 17.60±2.00; p=0.042). To confirm that the polymorphic TLR4 variant was transcribed, we investigated TLR4 mRNA by real-time PCR using specific Hybprobes designed for genotyping, which were 100% specific for the wild-type gene, but possessed one mismatch for the polymorphic TLR4 variant. The LightCycler-performed melting point analysis of the products revealed that indeed for carriers of the heterozygous polymorphisms both the wild-type (melting point 61 °C) and the polymorphic mRNA variant (melting point 56 °C) are transcribed in equal shares (figure 4). To investigate whether the LPS-binding capacity of monocytes from homozygous wild-type subjects is different from heterozygous subjects, we performed a FACS analysis. Monocytes from 37 homozygous wild-type subjects and monocytes from 18 heterozygous subjects were incubated with 0.35 ng/ml FCHSE-labelled LPS. We observed similar LPS-binding to monocytes of both groups (median of relative fluorescence: (+\+): 35.70±1.74 vs. (+\): 30.91±1.12).

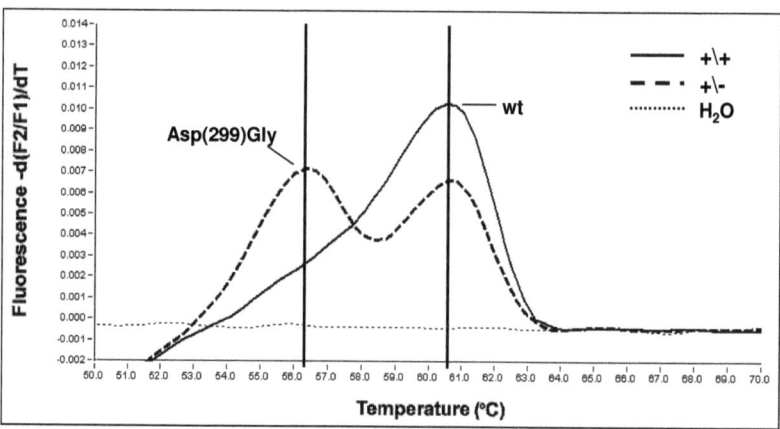

Figure 4 Carriers of the heterozygous Asp(299)Gly TLR4 polymorphism transcribe the wild-type and the polymorphic TLR4 variant RNA was prepared, reversely transcribed into cDNA and quantified by real-time PCR. The presence of the TLR4 wild-type (melting point 61 °C) and/or TLR4 Asp(299)Gly polymorphic variant (melting point 56 °C) was analysed by melting point analysis. TLR4 +/+ indicates the wild-type and TLR4 +\- the heterozygous polymorphic genotype.

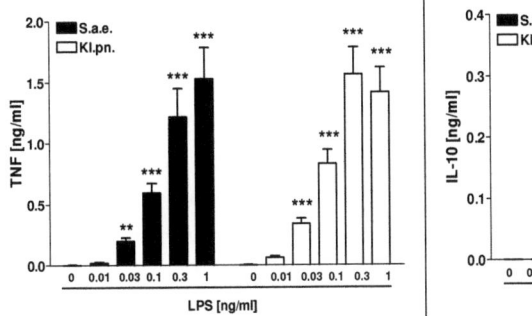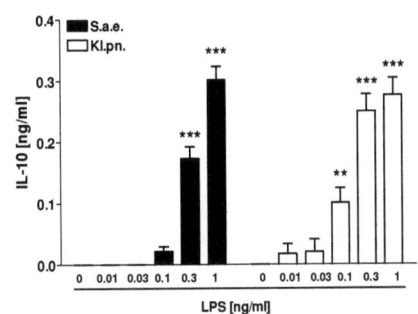

Figure 5 Induction of significant amounts of IL-10 from whole blood requires a higher LPS concentration than TNF induction One ml of 20% human whole blood from 24 healthy volunteers of the wildtype genotype was incubated in the presence of LPS at the concentrations indicated for 20h. TNF and IL-10 were determined in the cell-free supernatants by ELISA. Data are means ± SEM. *p < 0.05; **p < 0.01; ***p < 0.001 and indicate significance versus the control.

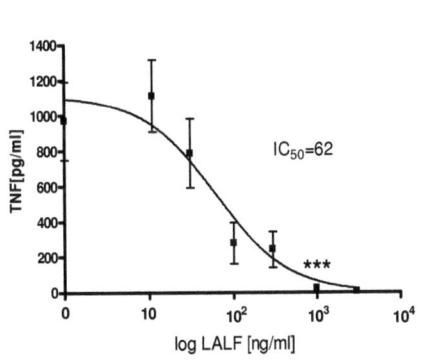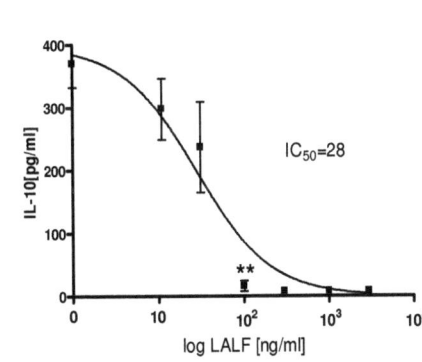

Figure 6 IL-10 induction is more sensitive to LPS inhibition than TNF One ml of 20% human whole blood from eight healthy volunteers was incubated in the presence of 1 ng/ml LPS or 1 ng/ml LPS together with 10 ng/ml-10 µg/ml LALF for 20h. TNF and IL-10 were determined in the cell-free supernatants by ELISA. Data are means ± SEM. **p < 0.01; ***p < 0.001.

Since it must be assumed that the polymorphic TLR4 variant, although not being defective in LPS binding, is defective in LPS-responsiveness, we compared the sensitivity of TNF- and IL-10 release to the concentration of LPS used for stimulation. When a concentration response curve ranging from 10 pg/ml to 1 µg/ml LPS in whole blood from healthy volunteers of only the wild-type genotype was performed, we found that IL-10 release required significantly stronger LPS stimulation than TNF release (figure 5). To avoid bacterial species-specific results, these experiments were performed with LPS from two different enterobacterial strains (*S.a.e.* and *Kl.pn.*). The two LPS exhibited a similar potency to stimulate the release of TNF, IL-1β, IL-8, IL-10 and IFNγ at the highest concentration employed (shown for TNF and IL-10 in figure 3), but while 30 pg/ml of LPS from both strains resulted in a significant release of TNF, 100 pg/ml LPS (*Kl.pn.*) and 300 pg/ml LPS (*S.a.e.*) were necessary to induce significant IL-10 release. In line with this observation, LPS (1 ng/ml)-inducible IL-10 release was significantly more susceptible to inhibition of LPS by the LPS-neutralizing factor LALF (figure 6).

3.5 Discussion

The only experimentally proven effect of the Asp(299)Gly polymorphism of TLR4 on inflammatory responses is a diminished release of the anti-inflammatory cytokine IL-10 upon in vitro stimulation of blood leukocytes from subjects with heterozygous expression of the polymorphic TLR4 variant with LPS [119]. This previous observation was now reconfirmed using a new study collective. It was shown that it also translates to isolated monocytes and that a significant reduction of IL-10 formation in case of subjects with heterozygous TLR4 polymorphism already occurs at IL-10 mRNA level, while the TNF mRNA levels remained unaffected. This, first of all, argues against an artefact of multiple testing in the previous study [119], where IL-10 was only one parameter measured among many.

The aim of this study was to investigate the underlying mechanisms responsible for the selective effect of the Asp(299)Gly polymorphism of TLR4 on cytokine release, which could not be explained by a delay in IL-10 release. In vitro transfection experiments had proven that the polymorphic variant of TLR4 is non-functional for LPS signalling [64], and it is assumed that the mutation at position 299, which is located in the extracellular LRR region of the TLR4 receptor, results in modified LPS binding. This is also supported by a recent study which provides evidence that the TLR4 mutation affects interaction with receptor agonists or co-receptors rather than intracellular signalling [122]. Therefore, it seemed likely that polymorphism carriers might suffer from impaired LPS responsiveness. However, in our study, neither the total TLR4 surface expression nor the LPS-binding capacity of monocytes was reduced in heterozygous subjects, although it was confirmed that the polymorphic gene variant is transcribed. The latter finding suggests that some of the expressed TLR4 receptors would be aberrant and not responsive. The unaltered LPS binding capacity of monocytes could be explained by the assumption that LPS initially binds to MD-2 and is then presented to TLR4, what in case of the Asp(299)Gly variant would not result in TLR4 activation. The fact that subjects with a TLR4 polymorphism showed an increased TLR4 surface expression might even indicate that the expression of the polymorphic non-functional variant is partially compensated by a stronger expression of the wild-type TLR4.

Our results clearly indicate that IL-10 induction in general requires stronger LPS stimulation of monocytes than TNF induction and that IL-10 release is more susceptible to inhibition of LPS by a neutralizing agent. Divergent sensitivities of the TNF and IL-10 ELISA, which would influence these results, were excluded. Taken together, this means that significant IL-10 expression requires a higher density of activated receptor complexes than TNF and thus is more susceptible to a lack of functional receptors, like in the case of the TLR4 polymorphism. Although both TNF and IL-10 are released in response to TLR4 stimulation, the signal transduction pathways, which initiate gene transcription, differ. While pro-inflammatory cytokines like TNF are induced via a synergistic interplay of the NF-κB pathway and activation of the MAPK-kinases ERK1/2, JNK and p38, IL-10 induction is dependent on p38 and the transcription factor Sp1, but does not involve ERK1/2 and NF-κB [123-125], which may already explain why TNF induction is more sensitive to LPS stimulation than IL-10.

However, we investigated several key parameters known to be relevant for IL-10 induction including p38 on the basis of phosphorylated p38 by Western blot analysis, as well as the role of the MyD88 independent TRIF/IRF pathway [126], and the induction of cyclooxygenase-2 and PGE2 [127]. We observed no difference between homozygous wild-type and heterozygous polymorphic subjects in any of these experiments (unpublished data). For TNF induction it is believed that LPS binding to the TLR4 receptor complex alone is sufficient to induce TNF [128], though this has not been investigated for IL-10 so far. One might speculate that IL-10 induction requires further processes like internalization of the LPS/receptor complex and intracellular processing. Preliminary results obtained with LPS coated to surfaces support this hypothesis but were not finally conclusive.

Given the pivotal role of LPS and its receptor TLR4 in bacterial immune recognition, this study gives a first explanation of specific inflammatory alterations in heterozygous Asp(299)Gly polymorphic subjects. The resulting proinflammatory phenotype could hence be a risk factor for excessive inflammation. Consistent with this, the TLR4 Asp(299)Gly polymorphism has been convincingly linked with inflammatory bowel disease and ulcerative colitis [66], in which IL-10 reduction is known to play a decisive role [31].

Therefore, it would be of major interest to investigate LPS-inducible IL-10 levels in patients with inflammatory bowel disease carrying the Asp(299)Gly polymorphism.

4

LPS-inducible anti-inflammatory responses are not diminished in Crohn's disease patients with heterozygous Asp(299)Gly polymorphism

Oliver Dehus[1], Gerhard Rogler[2], Jochen Hampe[3], Stefan Schreiber[3,4] and Corinna Hermann[1]

[1]Biochemical Pharmacology, University of Konstanz, Germany;
[2]Division of Gastroenterology and Hepatology, Department of Internal Medicine, University Hospital of Zürich, Rämistrasse 100,8091 Zürich,Switzerland;
[3]Department for General Internal Medicine, Christian-Albrechts-University, Kiel, Germany;
[4]Institute of Clinical Molecular Biology, Christian-Albrechts-University, Kiel, Germany

Submitted

4.1 Abstract

Crohn's disease is an inflammatory bowel disease characterized by a relapsing or chronical inflammation of all layers of the intestinal wall. The toll-like receptor (TLR)4 Asp(299)Gly polymorphism, which is associated with reduced LPS-inducible IL-10 release in healthy volunteers, has been linked with inflammatory bowel diseases. We have investigated by incubations of human whole blood whether in Crohn's diseases patients LPS-inducible TNF or IL-10 release is influenced by a heterozygous TLR4 polymorphisms compared to patients with a homozygous TLR4 wild type phenotype. We found that neither TNF nor IL-10 release was significantly different between both patient groups, and was furthermore comparable to cytokine release levels of healthy volunteers, indicating that probably at this stage of disease deviations in cytokine release occur only at the inflamed mucosa and cannot be detected by stimulations of leukocytes taken from the peripheral blood.

4.2 Introduction

Crohn's disease is characterized by chronic mucosal inflammation, which is discussed to be a consequence of abnormal immune responses to the autologous intestinal flora and a dysbalance of proinflammatory and regulatory cytokines. Especially IL-10, which inhibits antigen presentation as well as release of proinflammatory cytokines, turned out to play a pivotal role in the pathogenesis of Crohn's disease. It was shown that individuals that are genetically predisposed to produce less IL-10 are at higher risk of developing inflammatory bowel disease [36, 37]. Furthermore, IL-10-deficient mice develop colitis after colonisation with otherwise non pathogenic Gram-positive bacteria suggesting that under these circumstances normal immunosupressive barriers are broken [129, 130]. This has been linked to a defect of TGF-β/Smad signalling in the IL-10-deficient mice, which prevents inhibition of toll-like receptor (TLR) 2 mediated proinflammatory gene expression [131]. Since about half of the enteric indigenous flora is made up from Gram-negative bacteria, defects in TLR4 signalling might as well be from importance. The single nucleotide polymorphisms (SNP) A(896)G of *tlr4*, which results in the exchange of the amino acid Asp by Gly at position 299, has been linked with Crohn's disease and ulcerative colitis [66, 67], where it was discussed to be associated with distinct clinical pictures [132]. Only recently, we have reported that the TLR4 Asp(299)Gly polymorphism is associated with reduced LPS-inducible IL-10 release in blood leukocytes from healthy volunteers, while the release of other proinflammatory cytokines is not affected [133]. In the present study we have investigated whether Crohn's disease patients with TLR4 Asp(299)Gly polymorphism show reduced LPS-inducible IL-10 release compared to patients with TLR4 wildtype genotype.

4.3 Material and Methods

4.3.1 TLR4 genotyping of Crohn's disease patients

The TLR4 Asp(299)Gly polymorphism was determined in a population of 63 Crohn's disease patients (sex ratio was 26 women to 37 men, median age 43, range 22-80) recruited at the University Hospital of Regensburg, Germany. The diagnosis of Crohn's disease was done on the basis of standard clinical criteria. Determination of the A(896)G *tlr4* SNP of the Crohn's disease patients was performed by SNPlex chemistry (Applied Biosystems, Foster City, CA, USA) on an automated platform with TECAN Freedom EVO and 384well TEMO liquid handling robots (TECAN, Männedorf, Switzerland) as described before [134].

4.3.2 Human whole blood incubation

Whole blood from 63 Crohn's disease patients and 30 healthy volunteers (all recruited at the University Hospital of Regensburg, Germany) was stimulated with 1 µg/ml LPS from *Salmonella abortus equi* (*S.a.e.*, Sigma). Incubations of whole blood and ELISA measurements were carried out as described [119]. All healthy volunteers were genotyped for the A(896)G *tlr4* SNP, in order to exclude carriers of the polymorphism, as previously described [133]. All experiments and measurements were carried out blindly with regard to the blood donors' genotypes.

4.3.3 Statistics

Statistical analysis was performed using the GraphPad Prism 4.0 program (GraphPad Software, San Diego, USA). Significance of differences was assessed by one-way ANOVA. Data are depicted as box and whiskers blots.

4.4 Results

To investigate whether Crohn's disease patients with TLR4 Asp(299)Gly polymorphism show reduced LPS-inducible IL-10 release in comparison to Crohn's disease patients with wild type genotype, 63 patients were genotyped and the inflammatory responses of their leukocytes were assessed by in vitro stimulation of whole blood. The genotyping revealed that 12 out of the 63 Crohn's disease patients were heterozygous carriers of the Asp(299)Gly TLR4 polymorphisms, while 51 possessed a homozygous wild type genotype. None of the patients was a carrier of a homozygous polymorphism. To compare the inflammatory capacity of leukocytes from Crohn's disease patients with and without heterozygous Asp(299)Gly TLR4 polymorphisms, LPS-inducible TNF and IL-10 release was assessed by incubations of human whole blood. A collective of 30 healthy volunteers with wild type genotype served as control group. As shown in figure 1, no significant difference in TNF or IL-10 release was detectable between Crohn's disease patients with heterozygous Asp(299)Gly TLR4 polymorphisms or homozygous wild type genotype. Furthermore, TNF and IL-10 release of both patient groups was not different from that of blood leukocytes of healthy volunteers with homozygous TLR4 wild type genotype.

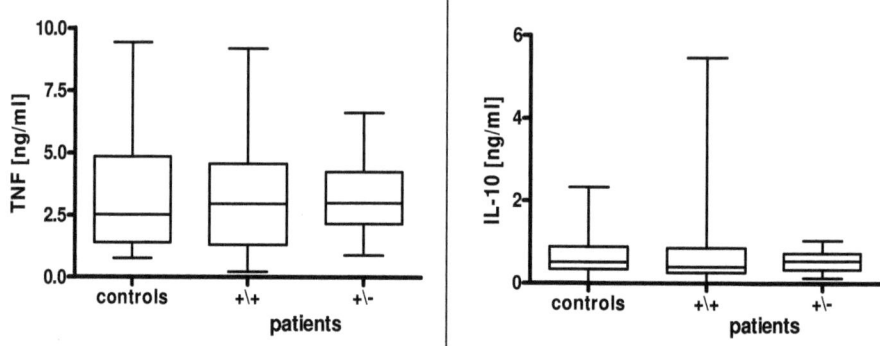

Figure 1 One ml of 20% human whole blood from 30 healthy volunteers with homozygous wild type genotype (controls), 51 Crohn's disease patients with homozygous wild type genotype (patients +/+) and 12 Crohn's disease with heterozygous polymorphic genotype (patients +/-) was incubated in the presence of 1 µg LPS for 20h. The release of TNF and IL-10 was determined in the cell-free supernatants by ELISA.

4.5 Discussion

The Asp(299)Gly polymorphism of the LPS receptor TLR4 has been linked with an increased susceptibility for inflammatory bowel disease in several studies [66, 67, 132, 135, 136] Since it was recently shown that in healthy subjects the same TLR4 polymorphism in heterozygous expression is associated with decreased LPS-inducible IL-10, while the release of other pro-inflammatory mediators remained unchanged [133], it was tempting to investigate whether this observation also translates to Crohn' disease patients, for which reduced anti-inflammatory responses are regarded as major risk factors for disease development. We have investigated the LPS-inducible responses of 63 Crohn's disease patients of which 51 possessed a homozygous wild type genotype and 12 a heterozygous polymorphic genotype. We found that LPS-inducible TNF and IL-10 release was comparable between both groups. Furthermore, the response of blood leukocytes from Crohn's disease patients was not significantly different from that of 30 healthy volunteers with wild type genotype. The association of decreased LPS-inducible IL-10 release and the Asp(299)Gly polymorphism of TLR4 had been confirmed in two independent study groups before, one consisting of 160 healthy volunteers of which 145 were carriers of a homozygous wild type genotype and 14 of a heterozygous Asp(299)Gly polymorphism [119] and a second one comprising 27 healthy volunteers, of which 17 carried a homozygous wild type and 10 a heterozygous polymorphic genotype [133]. The decrease in LPS-inducible IL-10 already occurred on m-RNA level and was supposed to be due to a lower number of responsive TLR4 receptors on monocytes of subjects with heterozygous Asp(299)Gly polymorphism [133]. Therefore, the comparability of LPS-inducible IL-10 levels between the Crohn's disease patients with and without TLR4 polymorphism was a surprise. We would have expected that the LPS inducible IL-10 levels are diminished in patients to the same extent or even stronger as it is the case for healthy subjects with heterozygous Asp(299)Gly polymorphism. Since several patients were treated with immunosuppressive medication, we cannot exclude that LPS-induced signal transduction is biased and deviations in inducible IL-10 release, which might have been observable without medication, are obliterated now. To check for modulatory effects due to immunosuppressive medication of the patients, we had included a group of healthy volunteers which had taken no medications and determined their LPS-inducible cytokine levels as well. Since especially TNF release, one of the most prominent proinflammatory mediators, was comparable between both patients groups and healthy controls, a dramatic influence of immune-modulatory medication can be excluded.

There is an increasing body of evidence that IL-10 is an important dampener of the ileal inflammatory processes in inflammatory bowel diseases. IL-10 concentrations for example are increased in the colon mucosa from diseased patients [137] and subjects that are genetically predisposed to produce less IL-10 are at higher risk for disease development [36, 37, 138]. We could not detect a reduced IL-10 release capacity of blood leukocytes from Crohn's disease patients with heterozygous Asp(299)Gly TLR4 polymorphism in our study. However, this was only a snap-shot taken from blood leukocytes of patients which had suffered from the disease for years and does not exclude that deviations in responses to LPS might exist which lead to significantly reduced IL-10 levels directly at the inflamed mucosal tissue of patients with Asp(299)Gly polymorphism. Therefore, it would be important to measure IL-10 concentration locally, best at the early onset of disease, to get non falsified results. If it turns out that patients with the Asp(299)Gly polymorphism are characterized by locally reduced IL-10 levels, these patients together with the group of genetically predisposed subjects might be the ideal target group that would benefit from locally administered IL-10 therapy.

4.6 Acknowledgements

The authors wish to thank all IBD patients, families and physicians for their cooperation. The cooperation of the German Crohn and Colitis Foundation (Deutsche Morbus Crohn und Colitis Vereinigung e.V.), the BMBF competence network "IBD", and of the contributing gastroenterologists is gratefully acknowledged. This study was supported by the German Ministry of Education and Research (BMBF) through the Competence Network IBD, National Genome Research Network (NGFN) and the popgen biobank. The project received infrastructure support through the DFG excellence cluster "Inflammation at Interfaces".

5

Endotoxin evaluation of eleven lipopolysaccharide by whole blood assay does not always correlate with Limulus Amebocyte Lysate assay

Oliver Dehus, Thomas Hartung[#] and Corinna Hermann

Biochemical Pharmacology, University of Konstanz, Konstanz, Germany
[#]ECVAM, EU Joint Research Centre, IHCP, Ispra, Italy

Journal of Endotoxin Research

5.1 Summary

More than 90% of all publications on endotoxin were carried out with endotoxins (lipopolysaccharide (LPS)) from enterobacteriaceae. We compared the immune stimulatory potency of eleven different LPS using human whole blood incubations. While the majority of LPS induced cytokine release equipotently, a thousand-fold more LPS from *Pseudomonas aeruginosa* or *Vibrio cholerae* was still less potent in inducing TNF, IL-1ß, IL-10 and IFNγ though it potently induced ng quantities IL-8. All LPS tested, regardless of the microorganism, showed toll-like receptor (TLR)4-dependence, except for the LPS from *P. aeruginosa* and *V. cholerae*, which were both TLR4 and TLR2 dependent. Interestingly, UV-inactivated *P. aeruginosa* bacteria, although Gram-negative, also showed TLR2- and TLR4-dependence. Re-purification of the commercial LPS preparation by phenol re-extraction led to a complete loss of the TLR2 dependency, indicating contaminations with lipoproteins. In the Limulus Amebocyte Lysate Assay, often performed to exclude contaminations in purified water likely to originate from *P. aeruginosa*, *P. aeruginosa* LPS was only two-fold less potent than LPS from *S. abortus equi* or the assay standard LPS from *E. coli*. This results in an overestimation of pyrogenic burden by a factor 500 in the sample when compared with the biological activity of highly purified *P. aeruginosae* LPS in human whole blood.

5.2 Introduction

Lipopolysaccharide (LPS), which makes up about 75% of the surface of Gram-negative bacteria, is known to be their major immune stimulatory principle (for review see [139]). It is released from the bacterial surface when the bacteria multiply, or when they die and lyse, leading to the activation of immune cells, as well as epithelial, endothelial or smooth muscle cells [140]. The recognition of LPS by host cells is an important step for the induction of inflammatory processes and anti-bacterial defense mechanisms, but might also lead to multi-organ failure and shock upon excessive systemic LPS exposure [141-143]. Chemical characterization and structural analysis of LPS of numerous enterobacteriaceae have revealed common structural features. The basic structure of LPS consists of a repetitive polysaccharide (O-antigen), which forms the outer part, the core oligosaccharide and the lipid A moiety, which is embedded in the outer membrane [77-79]. The O-antigen carbohydrate chain is a polymer of repeating oligosaccharides, which differ between species and determine the serological specificity of bacteria. In contrast, the structure of the lipid A, which consists of a phosphorylated disaccharide backbone, substituted with fatty acid, is highly conserved and exerts the endotoxic activity [80, 81, 144]. It is recognized by host immune cells via specific pattern recognition receptors, which immediately activate the host cells and stimulate cytokine release and complement activation leading to inflammatory responses [145]. The C3H/HeJ mouse has long been known to be hyporesponsive to LPS due to a spontaneous mutation of the *lps* gene [146]. Positional cloning revealed that the *lps* gene in these mice was the tlr4 gene with a point mutation [50]. At the same time some reports suggested TLR2 as LPS receptor [147, 148], but it soon became clear that the TLR2-dependent responses were induced by contaminating lipoproteins [149]. Although direct binding has not been demonstrated so far, it is believed today, that TLR4 together with MD-2 and the glycosylphosphatidylinositol anchored CD14 molecule confer sensitivity towards LPS [150]. Most of the initial studies have been done with LPS from enterobacteriaceae, since their structures were the first to be elucidated and synthesized chemically [81]. More recent studies, however, indicate that at least some LPS exist like the LPS from Bacteroides fragilis [151], Leptospira interrogans [85] or Porphyromonas gingivalis [86, 152] that are TLR2 dependent. During the last years, the structural and functional differences between LPS from different species became clear and the fact that the biological activity of LPS not only depends on the bacterial strain of origin but also on the cellular system used as the read-out (for review see [82, 83, 150]).

We therefore investigated the cytokine-inducing potency of eleven endotoxins from different bacterial species. As read-out system we chose the human whole blood incubation, which is likely to reflect the physiological situation, since all blood leukocytes are present in their physiological environment [153]. Further, we determined the TLR-dependence using cells from TLR2- and TLR4-defective mice.

5.3 Materials and Methods

5.3.1 Bacterial stimuli

5.3.1.1 Endotoxin

LPS from *Escherichia coli* K235, *Klebsiella pneumoniae*, *Salmonella abortus equi*, *Salmonella enteritidis*, *Salmonella typhimurium*, *Salmonella typhosa*, *Shigella flexneri*, *Serratia marcescens* 1A, *Vibrio cholerae* 569B (Inaba, O1), and *Pseudomonas aeruginosa* serotype 10 were purchased from Sigma, Deisenhofen, Germany. LPS from *Rhodobacter sphaeroides* was purchased from Quadratech, Epsom Surrey, England. LPS from *P. aeruginosa* and *S. abortus equi* were further purified by phenol re-extraction according to [149] to eliminate putative lipoprotein contaminations. Subsequently, dialysis of the LPS against aqua dest. was performed over night using a Spectra/Por MWCO 1 kDa membrane (Spectrum Laboratories, Inc., Ca, USA).

5.3.1.2 Bacteria

E. coli K-12 strain JM 109, a kind gift from Dr. Gerald Grütz, Charité Berlin, Germany, were grown in LB medium at 5% CO_2, 37°C, and stored at −80°C in physiological saline solution containing 25% glycerol. *P. aeruginosa* S10 (purchased from ATCC, Manassas, USA) were grown in LB medium at 5% CO_2, 37°C, and stored at −80°C in physiological saline solution containing 3.5% DMSO. Prior to incubation, the bacteria were washed twice with saline solution and UV-inactivated on ice for 60 min using an UV-Stratalinker 1800 (Stratagene, Jolla, CA, USA) at 9999 x 100 µJ.

5.3.2 Limulus Amoebocyte Lysate Assay

The kinetic Limulus Amoebocyte Lysate Assay (LAL, Charles River Laboratories, Sulzfeld, Germany, detection limit 0.1 EU/ml) was performed according to the manufacturer's protocol.

5.3.3 Human whole blood incubation

Human whole blood was taken from healthy volunteers using heparinized S-monovettes® (Sarstedt, Nürmbrecht, Germany) and diluted 1:5 with RPMI 1640 (Cambrex, Verviers, Belgium) containing 100 IU penicillin/100 µg streptomycin (Biochrom, Berlin, Germany) per ml in polypropylene reaction vials (Eppendorf, Hamburg, Germany). Stimulation was performed for 24h using sonified LPS or whole bacteria. After incubation at 37°C and 5% CO_2 in humidified air, the vials were shaken and centrifuged for 2 min at 400g. The cell-free supernatants were stored at -80°C until cytokine measurement by ELISA.

5.3.4 Isolation and stimulation of murine bone marrow cells

TLR4-mutated (TLR4$^{d/d}$) C3H/HeJ and respective wild type C3H/HeN mice were purchased from Charles River Laboratories (Sulzfeld, Germany). TLR2 knock-out (TLR2$^{-/-}$) mice were generated by homologous recombination by Deltagen (Menlo Park, CA, USA) and kindly provided by Tularik (South San Francisco, CA, USA). The animals were maintained under controlled conditions (22°C and 55% humidity, constant day/night cycle of 12h) and were fed ad libitum with Altromin 1314 (Lage, Germany). They received human care in accordance with the NIH guidelines as well with the legal requirements in Germany. To obtain bone marrow cells, mice were put under terminal pentobarbital anesthesia (Narcoren, Merial, Halbergmoos, Germany). Cells were isolated from both femurs by rinsing with 10 ml PBS and transferred into siliconized glass tubes (Vacutainer, BD Biosciences, Heidelberg, Germany). After centrifugation, cells were resuspended in RPMI 1640 (Cambrex) containing 10% FCS (PAA Laboratories GmbH, Pasching, Austria) and 100 IU penicillin/100 µg streptomycin (Biochrom) per ml and transferred to 96-well plates (Greiner, Frickenhausen, Germany) in a density of 5x 10^5 bone marrow cells/well. Cells where then stimulated with LPS or whole bacteria and incubated at 5% CO_2, 37°C for 24h. After incubation, the cell-free supernatants were stored at –80°C until cytokine measurement. The TLR2 agonist lipoteichoic acid (LTA), prepared in house from Staphylococcus aureus [75] and the TLR4 agonist LPS from *S. abortus equi* served as control stimuli in all experiments to ensure the responsiveness of the TLR-defective cells.

5.3.5 ELISA

Cytokines were determined by sandwich ELISA based on commercial antibody pairs against human TNF, IFNγ and IL-8 (Endogen, Perbio Science, Bonn, Germany), human IL-1β and IL-6 and murine TNF (R&D, Wiesbaden, Germany) and human IL-10 (Pharmingen Becton-Dickinson, Heidelberg, Germany). Binding of biotinylated antibody was quantified using streptavidin-peroxidase (Jackson Immuno Research, West Grove, PA, USA) and the substrate TMB (3,3',5,5'-tetramethylbenzidine, Sigma). Recombinant cytokines used as standards were obtained from the National Institute for Biological Standards and Controls, Herts, UK (TNF, IL-1β) and BD Biosciences, (IL-10, murine TNF, murine IL-6), PeproTech, Tebu, Frankfurt, Germany (IL-6 and IL-8) or Thomae, Biberach,

Germany (IFNγ). Assays were carried out in flat bottom, ultrasorbent 96-well plates (Nunc, Wiesbaden, Germany).

5.3.6 Statistics

Statistical analysis was performed using the GraphPad Prism 3.0 program (GraphPad Software, San Diego, USA). Significance of differences was assessed by the t-test for two groups and by one-way ANOVA followed by Bonferroni's post-test for experiments with more than two groups. In the figures *, ** and *** represent p values <0.05, <0.01 and <0.001, respectively.

5.4 Results

5.4.1 Immune-stimulatory potencies of eleven different LPS

The immune stimulatory potency of LPS from eleven different bacterial species, including LPS from the enterobacteriaceae *E. coli, S. abortus equi, S. enteritidis, S. typhimurium* and *S. typhosa*, as well as the enteropathogenic species *S. flexneri, S. marcescens* and *V. cholerae*, the opportunistic bacteria *K. pneumoniae* and *P. aeruginosa*, and the phototrophic bacterium *R. sphaeroides* were compared using human whole blood incubations.

For this purpose, concentration-response curves of the eleven LPS were performed and the release of the pro-inflammatory monokines TNF and IL-1β, the chemokine IL-8, the anti-inflammatory cytokine IL-10 and the lymphokine IFNγ was measured. The concentration-response curves of LPS from *E. coli, S. abortus equi, S.enteritidis, S. typhimurium, S. typhosa, S. flexneri, S. marcescens,* and *K. pneumoniae* were comparable for all cytokines measured and are exemplarily shown for *E. coli* LPS-induced TNF and IL-8 release in figure 1A and 1B, respectively. The minimal concentrations of these LPS necessary to induce significant cytokine release laid within one log-order for each cytokine and ranged from 0.01-0.1 ng/ml for the induction of IL-8, from 0.1–1 ng/ml for TNF, IL-1β and IFNγ, and were 1 ng/ml for IL-10 (table 1). In contrast, about a thousand-fold more LPS from *P. aeruginosa* and *V. cholerae*, i.e. 10-1000 ng/ml, were necessary to induce the release of significant amounts of TNF, IL-1β and IL-8, while no significant release of IL-10 and in case of *P. aeruginosa* also not of IFNγ could be induced.

Table 1 Minimal LPS concentration necessary for significant cytokine induction

LPS	cytokines				
	TNF	IL-1β	IL-8	IL-10	IFNγ
E. coli*	0.1	0.1	0.1	1	1
S. abortus equi *	1	0.1	0.1	1	1
S. enteridies *	0.1	0.1	0.01	1	0.1
S. typhimurium *	1	1	0.1	1	1
S. typhosa *	0.1	0.1	0.1	1	1
S. flexneri *	0.1	1	0.1	1	1
S. marescencs *	0.1	0.1	0.01	1	1
V. cholerae #	10	100	1000	-	100
K. pneumoniae *	0.1	1	0.1	1	0.1
P. aeruginosa #	1000	100	l00	-	-
R. sphaeroides #	-	100	100	-	-

Human whole blood from six volunteers was incubated with 0.1-1000 pg LPS/ml (bacterial species marked with *) or 0.1- 1000 ng LPS/ml (bacterial species marked with #) for 20h. Cytokines were determined in the cell-free supernatants by ELISA. The minimal LPS concentrations (ng/ml) that led to significant cytokine release compared to the unstimulated control ($p < 0.05$) was determined by Repeated Measures ANOVA followed by Bonferroni's Multiple Comparison test. Cytokine levels of unstimulated controls (pg/ml) were TNF, 92.2 ± 75.9; IL-1β, 6.9 ± 0; IL-8, 1065.6 ± 356.6; IL-10, 46.0 ± 29.1; IFNγ: 78.0 ± 49.3.

While the LPS from P. aeruginosa and V. cholerae, if applied in a thousand-fold higher concentrations, induced TNF of comparable amounts to E. coli LPS (figure 1A), they turned out to be very potent inducers of IL-8 (figure 1B). LPS from R. sphaeroides in high concentrations of 100 µg/ml induced the release of IL-1β and IL-8, but of no other cytokines measured. The main endogenous pyrogens are TNF, IL-1β and IL-6. On the one hand, TNF and IL-1β exert their pyrogenic potential through induction of IL-6, on the other hand, LPS can directly induce IL-6.

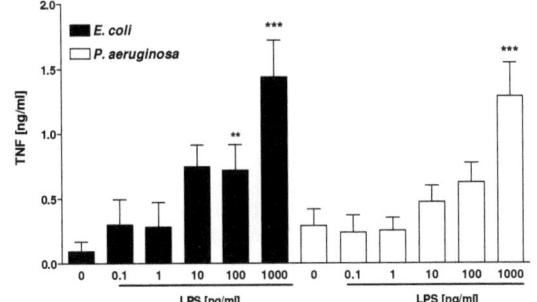

Figure 1 LPS concentration response curves One ml of 20% human whole blood was incubated with LPS from *E. coli* and *P. aeruginosa* at the concentrations indicated for 24h. TNF (A) and IL-8 (B) were determined in the cell-free supernatants by ELISA. Data are means ± SEM, n=6. ** and *** indicate significant cytokine release in comparison to the unstimulated control.

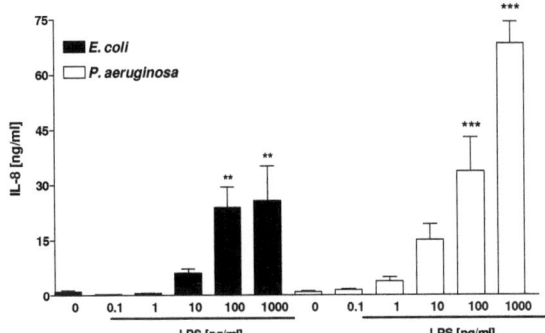

Therefore, the potency of *P. aeruginosa* LPS to induce IL-6 release has also been investigated. We found that in human whole blood, like for TNF and IL-1β thousand-fold more LPS from *P. aeruginosa* was necessary to induce IL-6 release in amounts comparable to LPS from enterobacteriaceae (IL-6 in ng/ml; LPS 10 ng/ml from S.a.e. 46 ± 5.3 versus LPS 10 µg/ml from P.a. 55 ± 6.3; n=6).

5.4.2 TLR-dependence of different LPS

In order to investigate the TLR-dependence of the different LPS, bone marrow cells from wild type and from TLR4-defective C3H/HeJ and TLR2$^{-/-}$ mice were stimulated with the eleven LPS and the release of IL-6 was measured. All LPS showed clear TLR4 dependence, except for the LPS from *P. aeruginosa* and *V. cholerae*, which were TLR4 and TLR2 dependent. These results are exemplarily shown for five of the eleven LPS in figure 2A and 2B.

A

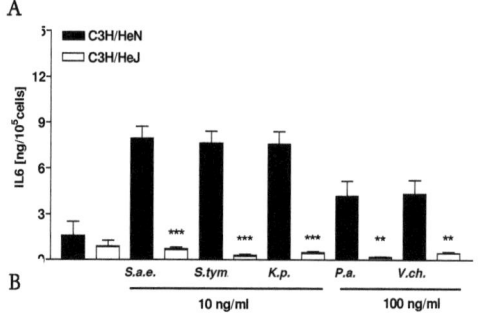

B

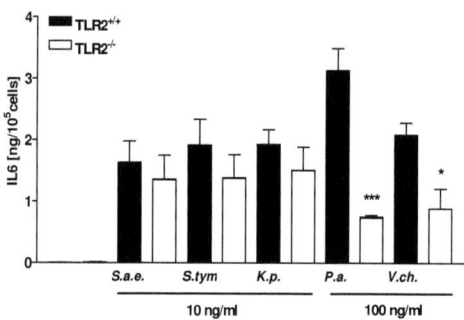

Figure 2 **TLR-dependence of different LPS** 5x 10^5 bone marrow cells from (A) C3H/HeN and C3H/HeJ and (B) $TLR2^{+/+}$ and $TLR2^{-/-}$ mice were incubated with LPS from *S. abortus equi* (S.a.e.), *S. typhimurium* (S.tym.), *K. pneumoniae* (K.p.), *P. aeruginosa* (P.a.) and *V. cholerae* (V.ch.) in the concentrations indicated for 24h. IL-6 was determined in the cell-free supernatants by ELISA. Data are means + SEM, n=4. *, **, *** indicate significant difference in IL-6 release in comparison to the cells from wild type mice.

5.4.3 Potency and TLR dependence of whole *P. aeruginosa* bacteria

In order to characterize the cytokine-inducing potency of whole *P. aeruginosa* bacteria, human whole blood was stimulated with different amounts of UV-inactivated *P. aeruginosa* in comparison to UV-inactivated *E. coli*. As shown in figure 3A, at least thousand-fold more *P. aeruginosa bacteria* than *E. coli* were necessary to induce comparable amounts of TNF, while the amount of maximal inducible IL-8 release was comparable (figure 3B). To investigate the TLR dependence, bone marrow cells from C3H/HeJ and TLR2-/- and their respective wild type mice were stimulated with 10^7/ml UV-inactivated P. aeruginosa or 10^5/ml UV-inactivated *E. coli*. Cytokine induction by *P. aeruginosa* was found to be strongly TLR4- and TLR2-dependent (IL-6 in ng/ml: C3H/HeN, 4 ± 0.7 vs. C3H/HeJ, 0.2 ± 0.03; TLR2+/+, 3 ± 0.4 vs. TLR2-/-, 0.4 ± 0.4; both n=8, p< 0.001), while *E. coli* induced cytokine release was only dependent on TLR4 (IL-6 in ng/ml: C3H/HeN, 2 ± 0.6 vs. C3H/HeJ, 0.8 ± 0.02, p< 0.001; TLR2+/+, 0.7 ± 0.3 vs. TLR2-/-, 0.6 ± 0.03, both n=4).

5.4.4 Phenol re-extraction and pyrogenicity of *P. aeruginosa* LPS

To clarify whether the TLR2 dependence of the LPS from P. aeruginosa was due to lipoprotein contamination, phenol re-extraction of the commercial LPS preparation was performed and the cytokine-inducing potency of the re-purified LPS as well as the TLR-dependence was investigated. The phenol re-extraction led to a significant reduction of the cytokine inducing capacity of the LPS in concentrations ≥1 µg/ml, as indicated in figure 4 for TNF release. The same effect was observed for all other cytokine (data not shown). However, when commercial LPS from *S. abortus* equi was repurified, no significant loss of potency was observed (1 µg/ml LPS before vs. after repurification, TNF ng/ml: 4.4 ± 0.3 vs. 5.4 ± 0.2; n=7).

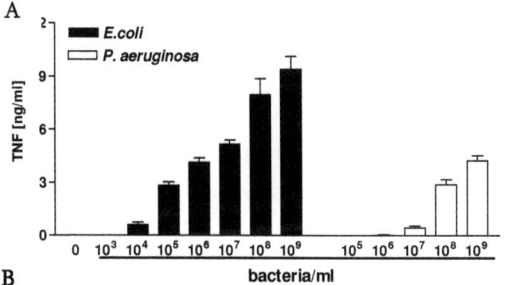
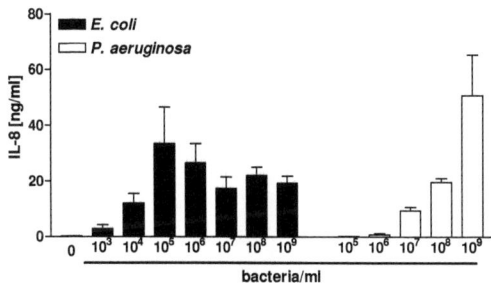

Figure 3 Comparison of cytokine induction by whole *P. aeruginosa* and *E. coli* One ml of 20% human whole blood was incubated with whole UV-inactivated *P. aeruginosa* or *E. coli* in the concentrations indicated for 24h. (A) TNF and (B) IL-8 were determined in the cell-free supernatants by ELISA. Data are means + SEM, n=4.

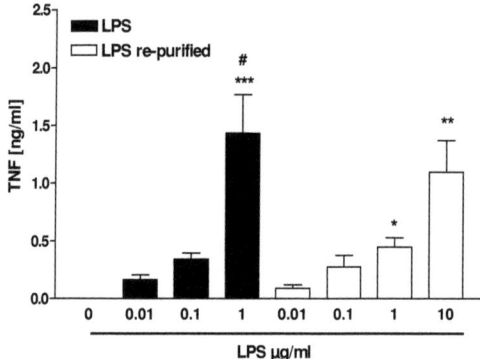

Figure 4 Comparison of cytokine induction by *P. aeruginosa* LPS before and after phenol re-extraction One ml of 20% human whole blood was incubated with LPS from *P. aeruginosa* before and after phenol re-extraction in the concentrations indicated for 24h. TNF was determined in the cell-free supernatants by ELISA. Data are means + SEM, n=4. *, ** and *** indicate significant TNF release in comparison to the unstimulated control, # indicates significant difference in TNF release in comparison to 1 µg/ml LPS after phenol re-extraction.

The stimulation of bone marrow cells from mice defective in TLR4 or TLR2 revealed that the re-purified LPS was still TLR4-dependent (figure 5A), but had lost its TLR2-dependent portion (figure 5B. Like in the human system, the re-purified P. aeruginosa LPS was less potent in the murine cells.

Determination of the endotoxic activity of *P. aeruginosa* and *S. abortus equi* LPS in the LAL showed that 1.7 ng *P. aeruginosa* LPS and 2 ng of the repurified *P. aeruginosa* LPS sufficed to achieve the same activity as 1 ng LPS from *S. abortus equi* or the LPS from *E. coli* O111, which served as reference material in the assay.

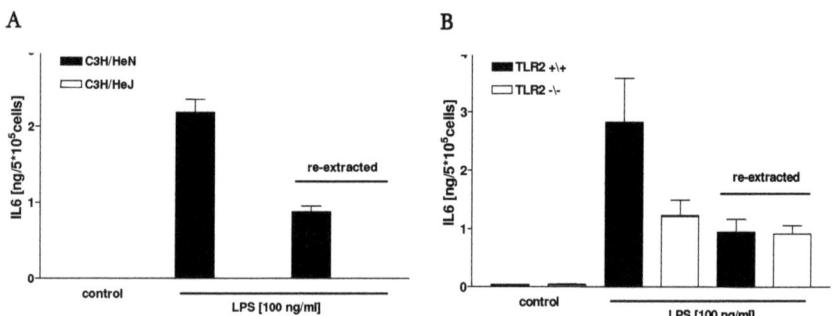

Figure 5 TLR-dependence of P. aeruginosa LPS before and after phenol re-extraction (A) 5x 10^5 bone marrow cells from C3H/HeN and C3H/HeJ and (B) TLR2$^{+/+}$ and TLR2$^{-/-}$ were incubated with LPS from P. aeruginosa before and after phenol re-extraction in the concentrations indicated for 24h. IL-6 was determined in the cell-free supernatants by ELISA. Data are means + SEM, n=4.

5.5 Discussion

Most of our knowledge on the immune stimulatory potency of endotoxins stems from studies performed with LPS from enterobacteriaceae. Only during the last years, functional differences of LPS from different bacterial species became evident, which are related to variations in LPS architecture [82, 83, 150]. It became evident that the endotoxic activity of a given LPS also crucially depends on the cellular system used as read-out, meaning that cytokine induction in human cells can be qualitatively different from murine cells and must not be related to the activation of the Limulus coagulation cascade [83]. We have investigated the immune stimulatory potency and TLR-dependence of eleven LPS from different bacterial species. The capacity of the LPS to induce the release of a variety of pro- and anti-inflammatory cytokines from primary leukocytes was assessed in human whole blood incubations, which closely reflect the physiological situation in humans in vivo [154]. Surprisingly, we found that the immune stimulatory potential and the induced cytokine profiles of almost all LPS, although derived from different bacterial species, were remarkably similar. Except for the LPS from *P. aeruginosa* and *V. cholerae*, the minimal LPS concentration to induce significant cytokine release, laid within one log-order for each cytokine ranging from 0.01-1 ng LPS/ml. The immune stimulatory activity of LPS from *P. aeruginosa* and *V. cholerae* was about a thousand-fold weaker and even at these high concentrations not the whole spectrum of cytokines was induced. Both LPS failed to induce IL-10. It should be noted that also for the other, more potent LPS, a higher LPS concentration was necessary to obtain significant IL-10 release, compared to TNF, IL-1β or IL-8, indicating that apparently a stronger stimulation of the monocytes is necessary to trigger IL-10 release. LPS from *R. sphaeroides* in high concentrations of 100 µg/ml induced the release of IL-1β and IL-8, but of no other cytokines.

The different endotoxic potencies of LPS molecules have been attributed to their architecture, which despite a common basic structure, can vary considerably. The highest variability is found in the part distal from the bacterial surface, the O-chain of the LPS, less variability in the core structure, but even the conserved lipid A, the structure that is recognized by innate immune receptors, shows variations in the acyl chain length, substitution patterns of the phosphates and the nature of fatty acids [83]. For full toxic activity in vivo and in vitro, a lipid A, consisting of a β-1,6-linked D-glucosamine disaccharide bearing two phosphate groups in position 1' and 4', substituted with six fatty acids 12 to 14 carbons in length, is required. Almost all deviations from this structure result in reduced endotoxic activity [77, 80].

The "optimal" lipid A structure described above is common for most enterobacteriaceae and enteropathogenic bacteria [79], i.e. in our study the LPS from *E. coli, S. abortus equi, S. enteritidis, S. typhimurium, S. typhosa, S. flexneri* and *S. marcescens*, as well as the opportunistic pathogen *K. pneumoniae*, which all exerted comparable potency in stimulating the release of various cytokines from whole blood. The LPS from *P. aeruginosa, V. cholerae* and *R. sphaeroides* also posse the di-phosphorylated β-1,6-linked D-glucosamine disaccharide, but differ in the fatty acid chain length and the type of hydroxylated and non hydroxylated fatty acid [79]. The LPS from *P. aeruginosa* contains only five fatty acids that are shorter in length (10 to 12 carbons) [150] and showed lower endotoxic activity in whole blood, which is in line with previous findings [155-157]. The O1 serotype LPS from *V. cholerae* possesses six fatty acids (12 to 16 carbons and traces of C18) [158] and was also a weaker cytokine inducing stimulus. The LPS from *R. sphaeroides* contains only five, partially unsaturated fatty acids [159], and has been shown before to act as a LPS antagonist on the level of ligand receptor interaction [84, 160, 161]. The classical hexa-acyl LPS structure from enterobacteriaceae, which is suggested to have a conical shape, seems to be the optimal structure for TLR4 activation, while the penta-acyl LPS with cylindrical shape, like the LPS from *P. gingivalis*, tends to engage TLR2 [82, 162]. In line with this, all LPS from enterobacteriaceae were TLR4-dependent, while the LPS from *P. aeruginosa* was both TLR4- and TLR2-dependent. In case of whole, UV-inactivated P. aeruginosa, the cytokine inducing activity depended on both TLR2 and TLR4. Other reports also indicate that *P. aeruginosa* triggers both TLR2 and TLR4, as shown for the activation of airway cells [163, 164] and host defence in mice [165, 166]. Since the TLR2-dependent activity could be separated by phenol re-extraction, this appears due to contaminating lipoproteins rather than being a property of the *P. aeruginosa* LPS itself, which turned out to be TLR4-dependent only. Furthermore, a previous publication has shown, that human TLR4 also discriminates between penta-acylated and hexa-acylated LPS, like it can be isolated from cystic fibrosis patients [167]. Obviously, for weak endotoxin, the immune stimulatory effects of lipoproteins seem to be more pronounced. In this light, studies employing crude *P. aeruginosa* LPS preparations with regard to TLR dependency should be reassessed [157].

P. aeruginosa bacteria are ubiquitously found in the environment. They colonize multiple niches and use many different compounds as energy source [168, 169]. Because of their ability to grow in water at 4°C, they are a major source of contamination in purified water, especially of dialysis solutions or water prepared for injection purposes [170, 171]. They play an important role as opportunistic bacteria for patients with AIDS and for neutropenic patients, for infections of burn wounds, for lung infections in patients with nosocomial pneumonia or cystic fibrosis and acute ulcerative keratitis [172]. So far, many virulence factors of P. aeruginosa like flagella, pili, LPS, proteases, exotoxins and exoenzymes have been shown to contribute to the clinic of infection [173-175]. In the present study we demonstrate that the leukocyte activating capacity of the P. aeruginosa bacterium, assessed as its cytokine release pattern, as well as its TLR-dependency, is reflected by its LPS. In vivo, P. aeruginosa infections are characterized by a massive influx of neutrophilic granulocytes [176, 177]. In line with this, we found that whole P. aeruginosa, like their LPS, are potent inducers of IL-8, a chemoattractant for neutrophils. We show that the similar 1000-fold weaker inducing capacity for other cytokines of the P. aeruginosa LPS is also seen for the whole bacterium in comparison to E. coli, which fits with previous data for E. coli and P. aeruginosa obtained with the cell line MonoMac6 [155]. This observed weak endotoxic activity of P. aeruginosa might explain the fact that although P. aeruginosa shows a widespread occurrence in the environment, nearly all clinical cases are associated with immune compromised hosts and infection is rare in healthy people [178, 179]. Others have shown that for several clinical P. aeruginosa isolates, the potency of toxicity in vivo is accompanied by a stronger cytokine release capacity in vitro [180], which supports our hypothesis. Remarkably, the endotoxic potency of the P. aeruginosa LPS is strongly overestimated by LAL. While by LAL, highly purified P. aeruginosa LPS was half as potent as reference preparations from S. abortus equi or E. coli, it was a thousand-fold less potent with regard to cytokine release from human whole blood. A 10- to 100-fold weaker pyrogenicity in rabbits compared to other LPS like the ones from E. coli or S. typhimurium has also been reported [181]. This indicates that cytokine release from human whole blood, in contrast to the LAL, reflects the potency different LPS have in vivo. Given the fact that the vast majority of LAL testing is carried out on water samples, this discrepancy for the foremost water contaminant is of critical importance. This is another example that clearly indicates that the extent of activation of the Limulus cascade cannot be directly correlated with the pyrogenic potential for humans. In this case, the endotoxic properties are overestimated by the LAL, though more often, endotoxic activities, even of synthetic Lipid A analogs, are not detected by LAL [81, 170, 182].

5.6 Acknowledgement

We thank Dr. Sonja von Aulock for her help with the manuscript and Margarete Kreuer-Ullmann for excellent technical assistance. This study was supported by a fellowship from the Margarete von Wrangell program to Dr. Corinna Hermann.

6

Growth temperature induces switching of structural variants of *Listeria monocytogenes* lipoteichoic acid

Oliver Dehus[1], Markus Pfitzenmaier[2], Sarah Meier[3], Gunthard Stuebs[4], Christian Draing[1], Wilhelm Schwaeble[5], Siegfried Morath[6], Thomas Hartung[1,7], Armin Geyer[2] and Corinna Hermann[1]

[1]Biochemical Pharmacology, University of Konstanz, Konstanz, Germany; [2]Department of Chemistry, Philipps-University of Marburg, Marburg, Germany; [3]Deutsches Rheuma-Forschungszentrum, Berlin, Germany; [4]Institute for Microbiology and Hygiene, Charité, Berlin, Germany; [5]Department of Infection, Immunity & Inflammation, University of Leicester, Leicester, UK; [6]Joint Research Centre, IHCP, Ispra, Italy; [7]CAAT, Johns Hopkins University, Baltimore, USA

Submitted

6.1 Abstract

We investigated the expression of lipoteichoic acid (LTA) from *Listeria monocytogenes* and found two distinct structural variants of LTA (LTA1 and LTA2) with expression levels influenced by growth conditions. While both LTA consisted of a poly-glycerophosphate backbone bound via a disaccharide to a diacyl-glycerol moiety, one LTA type (LTA2) possessed a second phosphatidyl diacyl-glycerol moiety. As assessed by human whole blood incubations followed by ELISA for TNF, IL-1β, IL-6, IL-8 and IL-10, the cytokine inducing potential of LTA1 was comparable to that of other LTA, like LTA from *S. aureus* or *S. pneumoniae*. In contrast, LTA2 induced significantly less of the pro-inflammatory cytokines and failed to activate the L-ficolin dependent pathway of complement *in vitro*, while the induction of the anti-inflammatory cytokine IL-10 was comparable to LTA1. Most interestingly, changes in growth temperature induced a switch in the expression levels of LTA1 and LTA2 in the cell wall: while the amount of LTA1 was comparable, the expression of LTA2 was low when *Listeria* were grown at room temperature in broth (ratio of LTA1 to LTA2 was 1:0.06), but increased when *Listeria* were grown at 37 °C (ratio of LTA1 to LTA2 was 1:0.68). The observed shift in LTA expression, probably accompanying the switch from the saprophytic to the virulent state, indicates an important adaptation to the different structural requirements and, possibly, to enable immune evasion.

6.2 Introduction

The opportunistic pathogen *Listeria monocytogenes* can cause a severe food-borne disease known as listeriosis, mainly affecting immunocompromised individuals, pregnant women and their foetuses or newborns [109, 110]. After invasion of the host, *L. monocytogenes* is capable of escaping the phagolysosome, replicating in the cytosol, inducing rearrangement of the host cell cytoskeleton and spreading directly from cell to cell. These highly organized actions are mediated via the expression of several virulence genes. The best studied are encoded by the virulence regulon LIPI-1, which comprises among others the genes for listeriolysin O (*hly*), phospholipase C (*plcA,B*) and actin polymerase A (*actA*), and which is under direct control of the transcription factor PrfA [109, 113, 183, 184]. PrfA activity mediates the switch from the saprophyte to the pathogen and underlies complex, environment-dependent sensor mechanisms like temperature or carbon sources [109, 185-189]. *Listeria* are characterized by a facultative intracellular life cycle and can replicate outside as well as inside the host. *In vivo*, *L. monocytogenes* is exposed to the host's immune system during the initial phase of infection as well as during the ongoing infection, e.g. after host cell lysis. It is important to note that *Listeria* inducing the initial infection have replicated in an extracellular environment, while *Listeria* that are set free from infected cells within the host have grown and replicated in an intracellular, host cell-controlled environment.

Whereas numerous publications describe the influence of growth conditions on gene regulation, its influence on immune stimulatory structures like peptidoglycan, lipoproteins or lipoteichoic acid (LTA) [74, 190] has not been investigated. In particular LTA is a potent inducer of cytokine release from leukocytes [103, 191] and has also been shown to induce complement activation [98]. Furthermore, LTA is a basic component of the cell wall, and thus contributes to the homeostasis of physicochemical surface properties [92] and confers resistance to antimicrobial cationic molecules [94]. We have investigated the influence of different growth conditions, i.e. broth culture at room temperature (RT) or 37 °C, on the expression of LTA from *L. monocytogenes*. We show here for the first time, that differences in growth temperature induce a switch in the expression of distinct structural LTA variants, which furthermore differ with regard to their pro-inflammatory activities.

6.3 Material and Methods

6.3.1 Bacterial strains and cultivation

The *L. monocytogenes* wild type strains ATCC 43251, DSM 12464, EGD and the EGD derived PrfA-mutant strains EGD/$\Delta prfA$ (lacking PrfA, referred to as ΔPrfA in the following) and EGD/$\Delta prfA$/pERL3/prfA* (constitutively expressing PrfA, referred to as PrfA* in the following), kind gifts from Prof. A. Goebel (Biocenter for Microbiology, University of Würzburg, Würzburg, Germany), were cultivated by shaking at 150 rpm in brain-heart-infusion medium (BHI, BD Biosciences, Pharmingen, Heidelberg, Germany) at 37°C or RT under aerobic conditions. Bacteria were harvested during the late exponential phase and washed with PBS. Quantification was carried out via PCR.

6.3.2 PCR

Genomic DNA from 1x 10^9 *L. monocytogenes* was prepared using the DNeasy® Tissue Kit from Qiagen, Hilden, Germany, as standard. Quantitative real-time (qRT) PCR was carried out for listeriolysin O encoding *hly* (sense: 5´-CAT GGC ACC ACC AGC ATC T-3´, antisense: 5´-ATC CGC GTG TTT CTT TTC GA-3´ [192]). Primers were purchased from Thermo Hybaid, Ulm, Germany.

6.3.3 Intracellular culture of *Listeria monocytogenes*

The human monocytic celline THP-1 (clone 238, T. Jungi, Berlin, Charite, Germany) was cultured in 1000 ml-bioreactors containing a cell compartment of 15 ml (celline classic 1000, Integra Biosciences, Bern, Switzerland) in RPMI 1640 (12-702F/U1, Lonza, Verviers, Belgium) supplemented with 20% FCS at 37°C and 5% CO_2. The cells were removed for washing with RPMI 1640 and the medium was changed every 4 days. A final cultivation step of 10 days completed the cycle. At the density of 2-3x 10^7 THP-1 cells/ml, *L. monocytogenes* (ATCC 43251) harvested from the exponential growth phase were added with a MOI of 10. After 40 minutes, non-invasive extracellular bacteria were removed by washing. Gentamycin (100 µg/ml) was added to prevent extracellular proliferation. After 15-18h of incubation, the cells were washed with PBS. The host cells were lysed in ultrapure distilled water for 1 min. Subsequently, the bacteria were purified from cell debris and cytosolic components by discontinuous gradient centrifugation with layers of 60%, 30% and 20% of iodixanol (Optiprep, Sigma, Daisenhofen, Germany), spinning for 1 h at 4°C and 45 000 g. The *Listeria* were then washed in PBS and directly employed in further experiments or extraction of LTA.

6.3.4 LTA extraction

LTA was purified from *S. aureus* (DSM 20233), *S. pneumoniae* (R6) and *L. monocytogenes* (ATCC 43251, EGD, ΔprfA and PrfA*) cultivated under cell-free conditions either at RT or 37°C, or intracellularly in THP-1 cells by butanol/water extraction as described previously (22). After hydrophobic interaction chromatography (HIC), the LTA containing fractions were identified by the phosphomolybdenum-blue assay. In case of LTA extracted from intracellularly grown *Listeria*, aliquots of the fractions were incubated in human whole blood and the resulting IL-8 release was measured by ELISA. Endotoxin contaminations of more than 0.5 EU/mg LTA were excluded employing the Limulus Amoebocyte Lysate Assay (Charles River Laboratories Sulzfeld, Germany).

6.3.5 Structural analysis

NMR spectroscopy

LTA from ATCC 43251 and DSM 12464 were analyzed by ^{1}H and ^{13}C nuclear magnetic resonance (NMR) spectroscopy as described for other LTA [97, 99, 100]. All spectra were recorded on Bruker DRX500 (500 MHz) and AVANCE 600 (600 MHz) spectrometers at 300 K using 5 mm BBI probe heads and can be viewed as supplementary data. In brief, the LTA were dissolved in D_2O with sodium 3-trimethylsilyl-3,3,2,2-tetradeuteropropanoate (TSP-d_4) added as internal chemical shift reference for ^{1}H NMR (δ_H 0.00 ppm), and acetone for ^{13}C (δ_H 30.02 ppm), respectively. For ^{31}P NMR 2% phosphoric acid was used as external standard (δ_P 0.00 ppm). The amount of LTA in each 0.6 ml sample ranged between 4 and 11 mg. Homonuclear assignments were based on two-dimensional double-quantum-filtered correlation spectroscopy (DQF-COSY), total correlated spectroscopy (TOCSY) and rotational nuclear Overhauser effect spectroscopy (ROESY) experiments using presaturation for water suppression. TOCSY and ROESY experiments were performed in the phase-sensitive mode using mixing times of 100 ms in TOCSY and 200 ms spinlock for ROESY, respectively. ^{13}C chemical shift assignments were obtained from gradient-enhanced HSQC spectra. Data acquisition and processing were carried out using standard Bruker software. The average number of repeating units in the poly-glycerophosphate backbone and the percentage of substitution were determined by integration of the corresponding peak volumes in the ^{1}H NMR.

GC/MS analysis of fatty acids

1 mg of LTA was dissolved in 0.6 ml methanol and 1 M HCl, covered with 1 ml n-heptane and incubated in glass tubes at 85°C. After 7 h, the reaction mixture was shaken and the

organic phase was dried. The methyl-esters of the fatty acids gained by acidic transesterification of the LTA were redissolved in n-heptane, and 1.5 µl were injected into a GC/MS (6890 Series GC-System/5973 Mass Selective Detector, Hewlett Packard, Böblingen, Germany). For normalisation, the C15 carboxylic acid methyl-ester (Merck, Darmstadt, Germany) was used as internal standard. The samples were vaporized during a gradient from 50°C to 280°C with a heating rate of 2°C/min. The peaks of the resulting chromatogram were quantified by calculating the area under the curve and the MS scans of the peaks were matched with the NBS75K library (Hewlett Packard).

6.3.6 Whole blood incubation

Whole blood was obtained from healthy donors (University of Konstanz, Germany) aged between 20 and 40. Blood donation routinely took place between 9 and 11 am. Acute infections were excluded by differential blood cell count using an ABX Pentra 60 (ABX, Montpellier, France). Incubations were performed as described previously [153]. Briefly, human whole blood was diluted five-fold with RPMI 1640 and stimulated over night with LTA from *S. aureus, S. pneumoniae*, or LTA1 or LTA2 from *L. monocytogenes* or whole *L. monocytogenes* that were inactivated by UV radiation on ice for 5 min at 1 J/cm^2 (UV-Stratalinker 1800, Stratagene, Jolla, CA, USA). The cell-free supernatants were stored at -80°C until cytokine measurement by ELISA.

6.3.7 Cytokine ELISA

Cytokines were measured by sandwich-ELISA in flat-bottom ultrasorbant 96-well plates (Nunc, Langenselbold, Germany) using commercially available antibody pairs and recombinant standards. Monoclonal antibody pairs against human TNF and IL-8 were purchased from Endogen (Perbio Science, Bonn, Germany), against human IL-1β and IL-6 from R&D (Wiesbaden, Germany) and against human IL-10 from BD Biosciences (Pharmingen, Heidelberg, Germany). Recombinant standards for TNF and IL-1β were kind gifts from S. Poole (NIBSC, Herts, UK), rIL-8 from PeproTech (Tebu, Frankfurt, Germany), rIL-10 and rIL-6 from BD Biosciences. The secondary biotinylated antibodies were detected by horseradish-peroxidase-conjugated streptavidin (Biosource, Camarillo, CA, USA), and TMB (3,3',5,5'-tetramethylbenzidine, Sigma) was used as substrate. The reaction was stopped with 1 M H$_2$SO$_4$ and the absorption was measured in an ELISA reader at 450 nm with a reference wavelength of 690 nm.

6.3.8 Complement assays

L-Ficolin binding and C4 cleavage were assayed according to Lynch et al. [98]. Briefly, LTA was immobilised over night at 4°C on flat-bottom ultrasorbant 96-well plates (Nunc GmbH & Co KG) in coating buffer (15 mM Na_2CO_3, 35 mM $NaHCO_3$, pH 9.6). Residual binding sites were blocked using Tris-buffered saline (TBS, pH 7.4) supplemented with 0.1% (w/v) human serum albumin (HSA, Aventis Behring, Marburg, Germany).

For measurement of L-Ficolin binding, human serum samples from healthy volunteers were diluted in TBS containing 10 mM $CaCl_2$, 0.05% (v/v) Triton-X100, 0.1% (w/v) HSA, pH 7.4 and incubated in the LTA-coated 96-wells plates over night at 4°C. 1 µg/ml of the human L-ficolin specific monoclonal antibody GN5 (Hycult biotechnology, Sanbio, Beutelsbach, Germany) was added and incubated for 90 min at RT. A 1:10000 dilution of the polyclonal goat-anti-rabbit peroxidase coupled detection antibody (GARPOX, DIANOVA, Hamburg, Germany) was added and incubated for 90 min at RT. TMB was used as substrate and optical density was measured as described above.

For measurement of C4 cleavage, human serum samples from healthy volunteers were diluted in 20 mM Tris, 1 M NaCl, 10 mM $CaCl_2$, 0.05% (v/v) Triton-X-100, 0.1% (w/v) HSA, pH 7.4 and incubated in the LTA-coated 96-wells plates over night at 4°C. Human C4 purified from serum of healthy volunteers as described previously [193] was diluted in 4 mM barbital, 145 mM NaCl, 1 mM $MgCl_2$, 2 mM $CaCl_2$, pH 7.4, added and incubated for 90 min at 37°C. A 1:1000 dilution of an anti-C4c antibody (Quidel, San Diego, USA), which was conjugated with biotin according to standard procedures, was added and incubated for 90 min at RT. Washing between the incubation steps was carried out with TBS, 0.05% Tween20, 5 mM $CaCl_2$. After incubation with horseradish-peroxidase-conjugated streptavidin for 30 min at RT, TMB was used as substrate and optical density was measured as described above.

6.3.9 Statistics

Statistical analysis was performed using GraphPad Prism 5 (GraphPad Software, San Diego, USA). Data are shown as means + standard error of the mean (SEM). For comparison of two groups, the paired two-tailed t-test and for more than two groups the repeated measures one-way-ANOVA followed by Tukey-post tests were applied. A *p*-value <0.05 was considered significant.

6.4 Results

6.4.1 Influence of the growth temperature on *L. monocytogenes* LTA

To investigate the influence of the growth temperature on the structure of *L. monocytogenes* LTA, LTA was prepared from *Listeria* (strain ATCC 43251) cultivated in broth either at RT or at 37°C. According to the phosphate profiles of the HIC fractions (figure 1A and B), both culture conditions resulted in two distinct phosphate peaks indicating two distinct LTA. Peak 1, comprising the fractions 44 to 50, corresponded to the phosphate peak that had also been observed for LTA extractions from other bacteria like *S. aureus* or *S. pneumoniae* [74, 100], whereas the second peak from fraction 53 to 64, eluting at more hydrophobic conditions, emerged here for the first time. The pooled LTA from these two peaks are referred to as LTA1 (peak 1) and LTA2 (peak 2). In contrast to LTA1, the dry weight yield of LTA2 differed according to the culture conditions: The ratio of LTA2 to LTA1 was more than ten times higher for *Listeria* cultivated at 37°C, i.e. 1:0.68 ± 0.034 for LTA1:LTA2 for growth at 37°C and 1:0.06 ± 0.012 for LTA1:LTA2 for growth at RT (data are mean ± SEM, n=3 experiments).

To determine whether intracellular growth further affects LTA composition, intracellular *Listeria* cultures in monocytic THP-1 cells were established in bioreactors at 37°C, and LTA was extracted from purified bacteria. Since the intracellular cultures allowed only limited yields of bacteria, only small amounts of LTA could be extracted, which did not allow establishing a phosphate profile. Therefore the fractions containing LTA could only be identified by cytokine induction in whole blood incubations, which is a more sensitive approach (22). As shown in figure 2, the resulting IL-8 profile again resulted in two peaks, arguing for the presence of LTA1 and LTA2. Determination of the dry weight indicated a LTA1:LTA2 ratio that was comparable to *Listeria* cultured at 37°C in broth.

To investigate whether the transcription factor PrfA is involved in the regulation of the relative expression levels of LTA1 to LTA2, we isolated LTA from the *Listeria* mutant strain PrfA*, which constitutively expresses PrfA, and from the *prfA*-deletion mutant ΔPrfA. It was expected that the PrfA* mutant would reflect wild type *Listeria* cultured at 37°C, where PrfA is highly induced, while the ΔPrfA mutant would reflect RT culture conditions, where PrfA is not expressed. However, the PrfA* mutant also showed a low LTA2 expression at RT whereas the ΔPrfA mutant was able to increase LTA2 expression at 37°C (data not shown). Control experiments were performed to guarantee that the corresponding wild type strain (EGD) showed a similar temperature dependent expression of LTA1 and LTA2 as the strain ATCC 43251 used for the initial experiments.

A

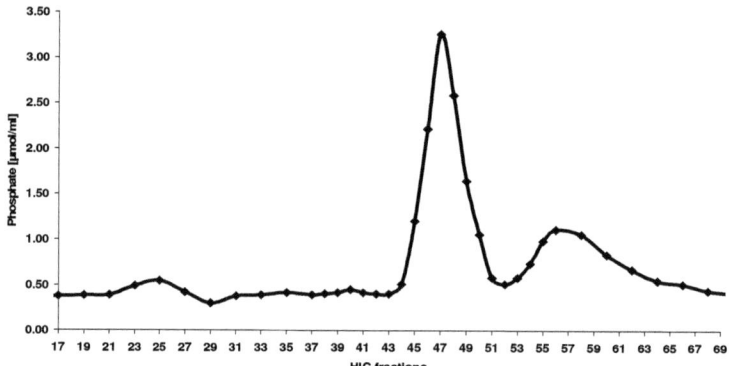

B

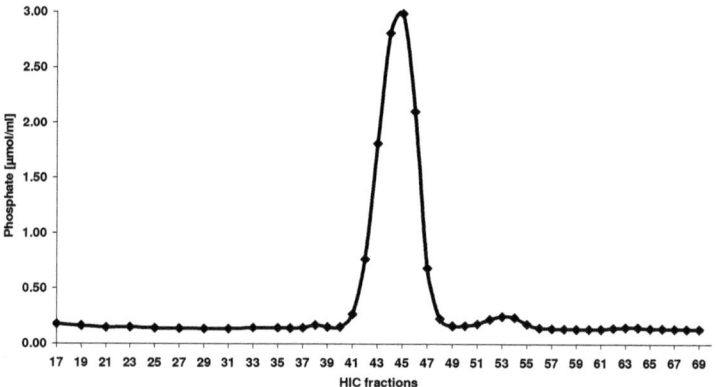

Figure 1: LTA preparation from *L. monocytogenes* results in two distinct LTA peaks
LTA was prepared from *L. monocytogenes* (ATCC 43251) cultivated under cell-free conditions at 37 °C (A) or RT (B) by butanol/water extraction. The water phase was separated via HIC and the phosphate content of the fractions was determined by the phosphomolybdenum-blue assay. Both profiles are representative for a minimum of three extractions each.

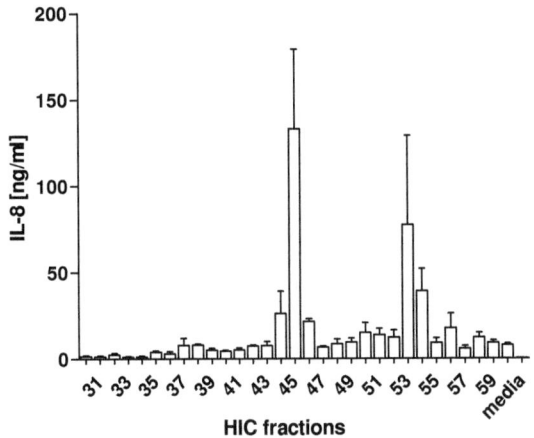

Figure 2: Cytokine profile from HIC fractions
LTA was prepared from intracellularly grown *L. monocytogenes* (ATCC 43251) by butanol/water extraction. The water phase was separated via HIC and the resulting fractions were examined for their IL-8 inducing properties by whole blood incubations followed by ELISA. Data are means + SEM, n=3.

6.4.2 Structural analysis of *Listeria* LTA

For structural analysis, both LTA were subjected to NMR and MS analysis. The results are depicted in figure 3 and table 1, respectively. For LTA1, a poly-glycerophosphate chain with an average of n=23 units was calculated. About 7% were substituted with α–galactose and 57% with D-alanine. The membrane anchor was composed of a 3(1)-(2´-O-α-D-galactopyranosyl)-α–D-glucopyranosyl]-1(3),2-diacyl-glycerol. The poly-glycerophosphate backbone of LTA2 was found to be approximately 50% shorter: an average of n=10 units substituted with 8% α–galactose and 53% D-alanine was calculated. The membrane anchor showed a rare second diacyl-glycerol moiety: 3(1)[6´-phosphatidyl-2´-O-(α-D-galactopyranosyl)-α–D-glucopyranosyl]-1(3),2-diacyl-glycerol. The acyl chains could be characterized as saturated and showed linear, iso- and antiso- branched methyl groups. To examine the structural differences of the acyl chains of LTA1 and LTA2 in greater detail, GC/MS analysis of the carboxylic acid-methyl esters gained by acidic transesterification of LTA1 and LTA2 was performed. In both cases, acyl chain lengths of 14, 16 and 18 were detected. The C18 acyl chains were unbranched, whereas the C16 acyl chains were partly branched with a methyl group on position 14, and the C14 chains carried a methyl group on either position 12 or 9. Taken together it turned out that in LTA1 the major portion of acyl chains (≅ 80%) consisted of branched, i.e. methylated, C14 and C16 chains, while in LTA2 the proportions of branched and linear acyl chains were more balanced (table I). Similar results were also obtained for LTA prepared from a second *L. monocytogenes* strain (DSM 12464, data not shown).

6.4.3 The pro-inflammatory activity of *Listeria* LTA1 and LTA2

To characterize the pro-inflammatory activity of LTA1 and LTA2 from *L. monocytogenes*, we examined cytokine induction in human whole blood as well as complement activation. Concentration response curves of LTA1 and LTA2 were compared to LTA prepared from *S. aureus* and *S. pneumoniae*. While cytokine release induced by LTA1 was comparable to that induced by LTA from *S. aureus* (figure 4) and *S. pneumoniae* (data not shown), LTA2 showed a shift to the right in its concentration response curve by a factor of ≥10 for the induction of the pro-inflammatory cytokines IL-6, IL-1β, TNF and IL-8, while its potency to induce the anti-inflammatory cytokine IL-10 was comparable to that of LTA1. We further investigated whether LTA2 can act as a competitive inhibitor of the cytokine release induced by LTA1: neither 1 µg nor 10 µg of LTA2 added to 1 µg LTA1 exhibited an inhibitory effect (data not shown).

We have described earlier that the lectin pathway of the human complement system is activated by LTA from *S. aureus* and *S. pneumonia* via binding to L-ficolin, which results in MASP-2 recruitment and activation of the cleavage of the zymogene C4 [98, 194]. In case of listerial LTA, LTA2 failed, even in the presence of high concentrations of human serum, to induce detectable levels of cleaved C4 in the C4 cleavage assay, while LTA1 induced C4 cleavage in a manner comparable to LTA from *S. pneumoniae* (figure 5A) and *S. aureus* (data not shown). This finding was accompanied by a thirty fold-reduced L-ficolin binding by LTA2 compared to LTA1 (figure 5B).

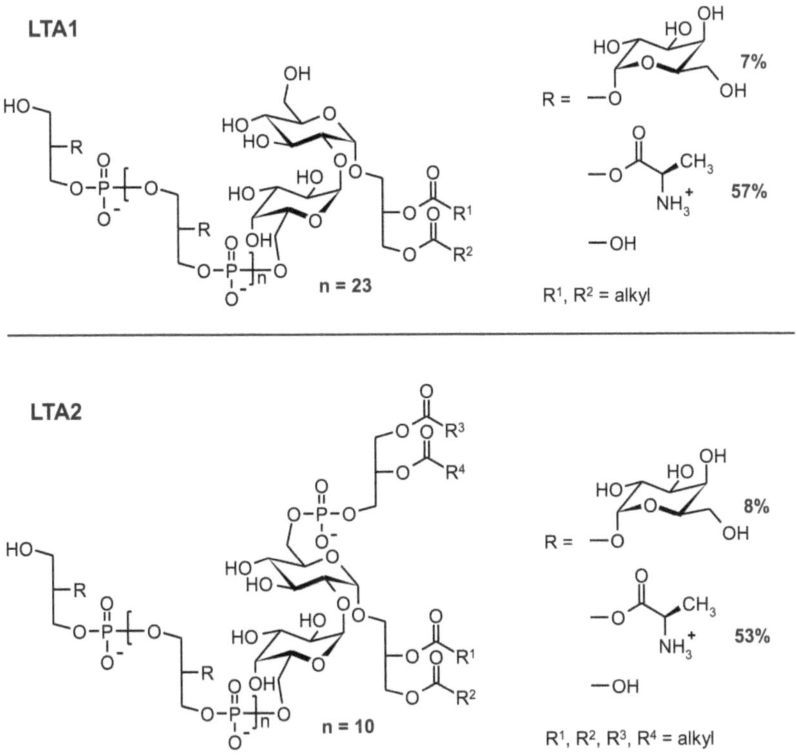

Figure 3: Structure and substitution pattern of LTA1 and LTA2
The percentage of alanylation, galactosylation and the average (av) number of the repeating units was determined from the ^{1}H NMR spectrum

Table I: Carboxylic acids of LTA1 and LTA2
Fatty acids of LTA1 and LTA2 were cleaved off by acetous transesterification and the resulting carbocylic-methyl esters were analysed via GC/MS. The chain lengths and the positions of natural methyl groups (ME) are given in the middle column (as resulted from MS analysis). The proportions of the carbocylic-methyl esters were determined by calculating the mean area under the curve from three GC chromatograms. Additionally, in the two lower lines branched versus linear carboxylic chains are given in their cummulative proportions.

LTA1 %	carboxylic-methyl ester	LTA2 %
9.42	C18	28.87
6.99	C16	24.02
32.24	C16 ME (14)	17.09
39.51	C14 ME (12)	22.86
11.84	C14 ME (9)	7.16
16.4	Sum of linear chains	52.9
83.6	Sum of branched chains	47.1

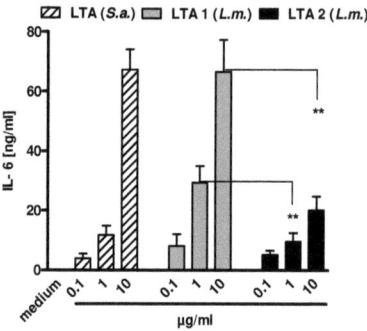
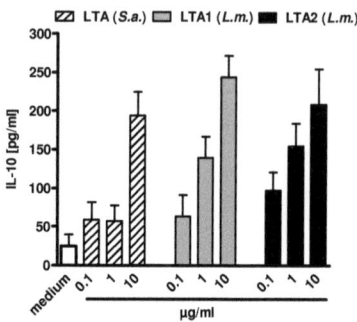
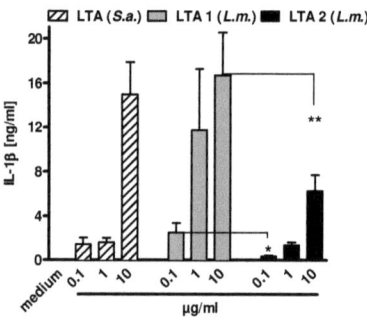
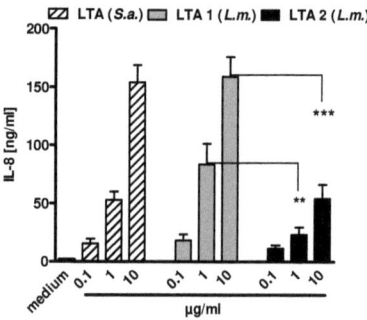
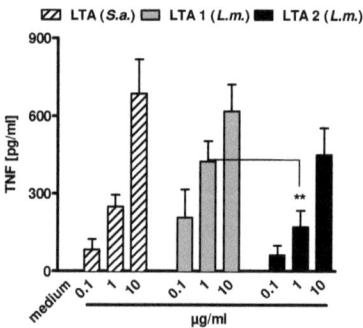

Figure 4: LTA2 possesses less pro-inflammatory potency compared to LTA1
Whole blood was incubated for 24h with LTA from *S. aureus* (DSM 20233) or with LTA1 or LTA2 from *L. monocytogenes* (ATCC 43251), all cultivated under cell-free conditions at 37°C, at the concentrations indicated. Release of cytokines was measured by ELISA. Data are means + SEM, n=9. Statistical analysis includes LTA1 and LTA2. * $p<0.05$, ** $p<0.01$, *** $p<0.001$

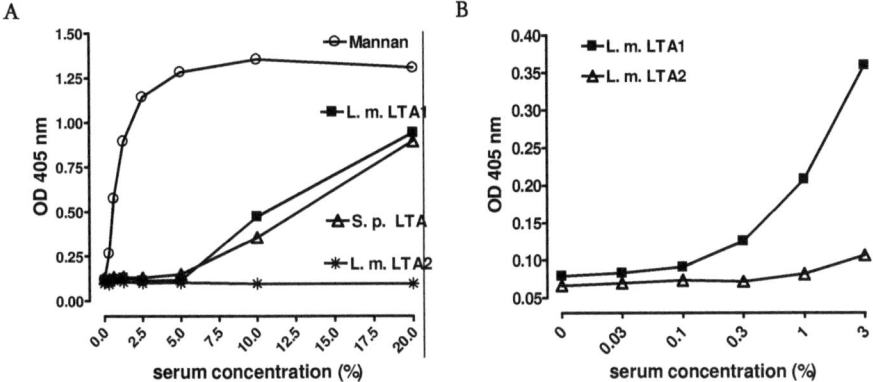

Figure 5: LTA2 fails to induce complement activation
A: Plates were coated with mannan, LTA1 or LTA2 from *L. monocytogenes* (ATCC 43251) or LTA from *S. pneumoniae* (R6), all grown at cell-free conditions at 37°C and incubated with human C4 in the presence of human serum in the concentrations indicated. The cleaving product C4c was quantified by ELISA. B: Plates were coated with LTA1 and LTA2 from *L. monocytogenes* (ATCC 43251) cultivated at cell-free conditions at 37°C and incubated with human serum in the concentrations indicated. L-ficolin binding was quantified by ELISA.
Each data set given is representative for at least three similar experiments.

6.4.4 Influence of different growth temperatures on the pro-inflammatory activity of whole *L. monocytogenes*

Having shown that LTA2 possesses a lower pro-inflammatory capacity than LTA1, we compared the cytokine inducing capacity of whole bacteria grown in broth either at RT or 37°C or intracellularly in THP-1 cells at 37°C (figure 6). For that purpose, human whole blood was incubated with different amounts of freshly harvested, UV-inactivated bacteria, and the release of IL-6, IL-1β, TNF, IL-8, and IL-10 was measured. The release of all mediators with the exception of IL-1β tended to be reduced when bacteria had been grown at 37°C in broth compared to RT. Control experiments confirmed that the cytokine inducing activity of *Listeria* cultivated in broth was not dependent on the growth phase. However, when intracellularly cultured *Listeria* were used to stimulate human whole blood, we found a significant reduction in the release of all pro-inflammatory mediators, while the induction of the anti-inflammatory cytokine IL-10 was comparable to *Listeria* grown at 37°C in broth.

6.5 Discussion

In general, the structure of LTA is highly conserved among a wide variety of species and studies with staphylococcal LTA mutants revealed that LTA plays an important role in colonisation, virulence and cell division [89, 90, 195, 196]. In *L. monocytogenes*, LTA is reported to function as a scaffold for the virulence protein internalin B (InlB), which mediates invasion of host cells [95, 96]. LTA is also a potent stimulus for the innate immune system, inducing the expression of a variety of inflammatory molecules [74, 99] and the L-ficolin dependent pathway of complement activation [98]. So far, we have isolated and examined LTA from several different bacteria including *S. aureus* [75, 99], *S. pneumoniae*, [100], *B. subtilis* [97] or *Lactobacilli species* [101], and observed only minor differences regarding immune stimulation, although some variations in structure, especially of the hydrophilic chains, existed [197]. In this work we present the novel finding that different culture conditions affect the LTA expression of *Listeria* strains. We identified two distinct structures of LTA, which differ in their pro-inflammatory activity while the ratio of LTA1 to LTA2 expression is influenced by growth temperature.

In comparison to LTA1, LTA2 was a weaker inducer of pro-inflammatory cytokine release and complement activation. Noteworthy, all incubations were performed on the basis of the weight (in µg quantities), which is approximately 1.5 times higher for LTA1 in comparison to LTA2. Therefore, LTA2 would be even weaker in comparison on a molar basis.

Determination of the LTA structure by NMR showed that the increased hydrophobicity of LTA2 as indicated by the elution profile was due to a shorter hydrophilic backbone compared to LTA1 and to the presence of a novel second phosphate-bound lipid anchor that was absent in LTA1. The building blocks of the sugar backbone were identical in LTA1 and LTA2, showing the same substituents and the same disaccharide moiety. Since the second diacyl-glycerol moiety of the lipid anchor of LTA2 is so far unique, it appears likely that it accounts for the altered inflammatory properties. As LTA2 does not antagonize immune stimulation by LTA1, sterical inhibition or the increased hydrophobicity of LTA2 may prevent it from binding to immune receptors. In the last years the cytokine inducing activity of LTA was questioned and was suggested to be due to contaminations with lipoproteins [198]. From our structural analysis we have no indications for lipoprotein contaminations and so far, no lipoproteins with inflammatory activity have been purified from *Listeria*. LTA extractions used to be performed with hot phenol, resulting in partly decomposed and often inactive material [74, 103]. Butanol/water extraction at ambient temperatures is gentler and generates biologically active material (22). However, this does not appear to be critical for the extraction of LTA from *Listeria*, as the structural analysis of our LTA is in agreement with a previous report of Uchikawa et al. [199] who used the hot phenol extraction method.

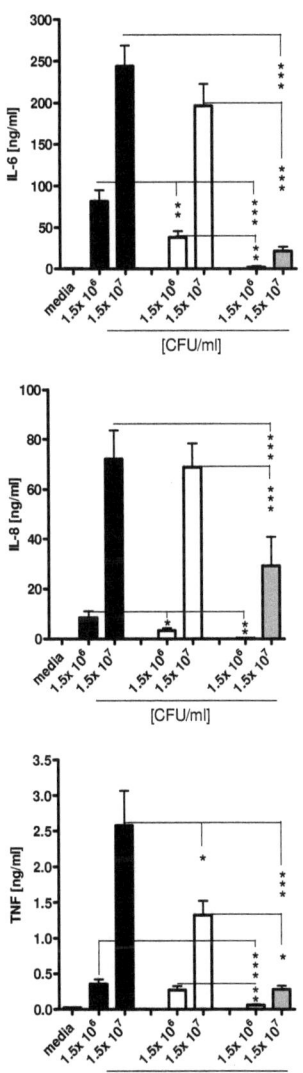
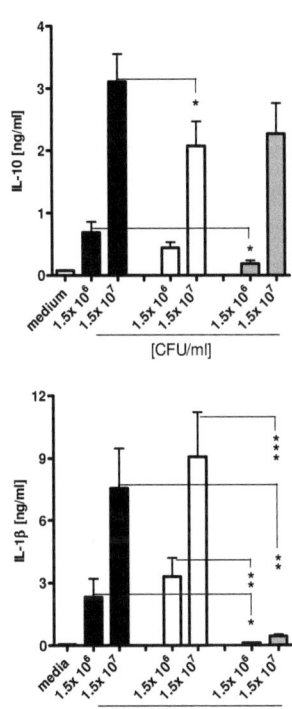

Figure 6: Growth conditions of *L. monocytogenes* affect cytokine release from whole blood
Whole blood was incubated with UV-killed *L. monocytogenes* (ATCC 43251), which had grown either cell-free at RT (black bars) or 37°C (white bars), or intracellularly in THP-1 cells (grey bars). Release of cytokines was measured by ELISA. Data are means + SEM, n=8 individual donors.
* $p<0.05$, ** $p<0.01$, *** $p<0.001$

Our findings indicate that switches in growth temperature from RT to 37°C already induce the switch to more LTA2 expression. This increased LTA2 expression was also seen for intracellularly cultured bacteria, although in this case the limited amount of extracted LTA did not allow further study of this phenomenon. It is well known that infection of the host activates a whole machinery of virulence factors, many of them under the control of the temperature sensitive transcription factor PrfA [113]. However, the use of *Listeria* mutants that either constitutively expressed PrfA* and thereby reflect the virulent state or which expressed no PrfA, reflecting the saprophytic state, clearly argued against a causal relationship between induction of PrfA expression and the increase in LTA2 synthesis. Nevertheless, the control of LTA2 expression by virulence factors cannot be negated in general: an involvement of the less well-examined VirR or of so far unknown players remains to be investigated [200].

Unfortunately, the pathways of LTA synthesis are not clarified sufficiently so far that the mechanism of temperature sensitive control could be elucidated. The different LTA structures could stem from decreased enzyme activity at 37°C or could be a product of specialized enzyme activity evolved for virulence. For *Listeria* it is hypothesised that the phosphatidyl-galactopyranosyl-glucopyranosyl-diacyl-glycerol is synthesized in the cytosol, involving a putative *S. aureus*-YpfP-homologue [201], and transported to the periplasm where the glycerol phosphate backbone polymerizes, involving a putative *S. aureus*-LtaS homologue [195]. A faster incorporation of the lipid anchor into the membrane due to the presence of four fatty acid residues in case of listerial LTA2 instead of only two in case of listerial LTA1 may explain the shorter backbone.

Although *Listeria* grown at 37°C in broth possessed a higher proportion of LTA2, i.e. the LTA structure with significantly lower pro-inflammatory activity, the inflammatory properties of the whole bacteria were comparable to *Listeria* cultured at RT. This result may be explained by the observation that the absolute amount of LTA1 was comparable under both culture conditions and that LTA2 did not show an antagonizing effect. This clearly indicates that the LTA1:LTA2 ratio is not the only driving force for cytokine induction. Interestingly, the pro-inflammatory properties of whole bacteria were significantly reduced when they were cultured intracellularly in THP-1 cells, which might be a further consequence of adaptation to the host cell environment. However, the intracellularly cultured bacteria were subjected to several manipulations like the influence of gentamycin, several purification steps or possible encapsulation by host cell proteins, which could not all be controlled [202, 203]. To exclude at least adverse effects of the purification procedure, we treated bacteria grown in broth with the same purification protocol as for intracellularly grown bacteria, and could thereby negate an effect on the inflammatory potential.

For *L. monocytogenes*, this is the first report of a temperature-inducible structural modification of LTA, an essential, highly conserved cell wall component, which might reflect adaptation to the host cell environment as well as a benefit for immune evasion.

6.6 Acknowledgements

The authors thank Prof. Dr. Werner Goebel and Regina Stoll for kindly providing the mutant *Listeria* strains. The excellent technical support by Leonardo Cobianchi is gratefully acknowledged.

7 Summarizing Discussion

The course and outcome of an infection is determined by the nature and virulence of the pathogen, the status of the host's immune system, the host's genetic and epi-genetic basis as well as the response on the treatment applied. The better we understand and control theses relations, the more adapted and successful will be the choices of prevention and therapeutic treatments.

The TLR4 polymorphism Asp(299)Gly

Of outstanding clinical relevance in this context are Gram-negative infections causing lethal septic shock or multi organ failure. The course of disease is hardly predictable and depends on the patients' susceptibility and on the effectiveness of his immune reactions [24, 64, 204]. These observations might be attributed to variances of genes involved in immune defence. The LPS receptor TLR4 represents one candidate of a polymorphic gene. *In vitro* transfection experiments revealed that the Asp(299)Gly polymorphism renders the receptor non functional [64]. Our own study regarding LPS binding and LPS-induced cytokine induction in primary human blood leukocytes from individuals with the TLR4 polymorphism belongs to the few experimental studies that investigate the mechanisms which might explain the association between the TLR4 polymorphism and increased susceptibilities towards infections or chronical inflammatory disorders. Initially, in LPS inhalation experiments the Asp(299)Gly polymorphism showed a blunted bronchoconstrictive response for some subjects with the Asp(299)Gly polymorphism [64], other investigations showed a higher prevalence of the Asp(299)Gly mutation in Gram-negative sepsis and a more severe course of disease [24, 205]. These first reports prompted several further studies, unfortunately with mostly controversial outcomes. Noteworthy, monitoring a huge collective for five years, the risk of developing carotid sclerosis was found attenuated in heterozygous individuals carrying the Asp(299)Gly polymorphism and proinflammatory serum proteins were reduced [206]. Taken together, these data suggest that the Asp(299)Gly TLR4 polymorphism might alter the initiation of inflammatory responses. Until now, only two functional studies are available, which report no influence of the TLR4 Asp(299)Gly polymorphism on inflammatory responses induced by Gram-negative stimuli [118, 119]. However, in one study from our laboratory, a decrease of LPS-induced IL-10 release from whole blood was reported [119]. In this study, more than twenty endpoints had been investigated in blood from 160 donors; therefore the physiological relevance of the statistic significance remained to be confirmed. Our follow-up study aimed to investigate the role of the Asp(299)Gly TLR4 polymorphism for LPS-inducible IL-10 release in greater detail and to clarify possible molecular mechanisms. One crucial finding was, in a new study collective, the confirmation of the decreased LPS-induced IL-10 release while TNF was not affected. IL-10 expression was found to be diminished in whole blood and in isolated monocytes already on the mRNA level. To elucidate this IL-10 specific effect, we investigated several key parameters known to be relevant for IL-10 induction including p38 [124], as well as the role of the MyD88 independent TRIF/IRF pathway [126], and the induction of cyclooxygenase-2 and PGE_2 [127]. We observed no difference between homozygous wild-type and heterozygous polymorphic subjects in any of these experiments (unpublished data). For TNF induction it is believed that LPS binding to the TLR4 receptor complex alone is sufficient to induce TNF [128], though this has not been investigated for IL-10 so far. One might speculate that IL-10 induction requires further processes like internalization of the LPS/receptor complex and intracellular processing. Preliminary results obtained with LPS coated to surfaces support this hypothesis but were not finally conclusive. However, our results clearly indicate that IL-10 induction in general requires stronger LPS stimulation of monocytes than TNF induction and that IL-10 release is more susceptible to inhibition of LPS by a neutralizing agent. This can explain, why the polymorphism specifically affects IL-10 expression, but the mechanistic key components remain to be clarified, by which the mutation of a receptor, whose activation initiates a variety of downstream events, ends up with the remarkably exclusive impact on one cytokine only. Taken together, the influence of the mutation surprisingly implies a proinflammatory phenotype, contrasting the majority of the references mentioned above concluding an increased susceptibility towards infections and arguing with impaired proinflammatory responsiveness. In contrast, particularly the associations with severe outcomes of sepsis might be defined much better by a reduced anti-inflammatory response, which means to reconsider therapeutic strategies. Importantly, as a risk factor for excessive inflammation, the TLR4 Asp(299)Gly polymorphism has also been convincingly linked with inflammatory bowel disease and ulcerative colitis [66], in which IL-10 reduction is known to play a decisive role [31]. Therefore, it was of major interest to investigate LPS-inducible IL-10 levels in patients with Crohn's disease carrying the Asp(299)Gly polymorphism, since this disorder is discussed to be a consequence of a dysbalance of proinflammatory and regulatory cytokines [36, 37]. We designed a study in order to investigate whether Crohn's disease patients with TLR4

Asp(299)Gly polymorphism show reduced LPS-inducible IL-10 release compared to patients with TLR4 wildtype genotype. Although the polymorphism was confirmed to be associated with Crohn's disease by our study, the phenotype of reduced LPS-induced IL-10 release was surprisingly not visible in the patients with a long-years history of this disorder. Also when the data were evaluated excluding subjects with stronger immune-modulating medication, no significant bias of the results was observed; neither was with a gender specific evaluation. We cannot exclude that the non-pathophysiological inter-individual variations of cytokine release, which we also observed in our groups of healthy individuals, account for that, but planning the study we estimated the collective to be of sufficient size. It has to be noted, that according to clinical standard criteria, the pathophysiology of Crohn's disease has a multifactorial and polygenic background. The TLR4 Asp(299)Gly polymorphism constitutes only one effector and an impact of the polymorphism possibly requires the presence of other factors like NOD2 polymorphisms (Hume IBD 08). Furthermore, this was only a snap-shot taken from blood leukocytes of patients which had suffered from the disease for years. It does not exclude that deviations in responses to LPS might exist which lead to significantly reduced IL-10 levels directly at the inflamed mucosal tissue of patients with Asp(299)Gly polymorphism. Furthermore, the impact of the polymorphism could be much more pronounced during the early development of the mucosal immune system, thus increasing the susceptibility rather than the outcome of Crohn's disease. Examining the local IL-10 concentration at the early onset of disease would be most advisable. If it turns out that patients with the Asp(299)Gly polymorphism are characterized by locally reduced IL-10 levels, these patients might be the ideal target group that would benefit from locally administered IL-10 therapy.

Non-classical bacterial immune-stimuli

Not only variations in immune receptors and signalling create an individual immune response, also the bacterial derived triggers account for the variety of pathogen-host interactions. Even a species-specific course of infection may comprise variable features depending on the nature of the pathogen. The toxicity of Gram- negative bacteria is defined by their LPS and knowledge about structural variations of the lipid A moiety is steadily increasing [76, 79]. Investigating the immune-stimulatory potential of eleven LPS in whole blood, we found that the resulting cytokine profiles of almost all LPS, although derived from different bacterial species, were remarkably similar and seem to be a common feature of enterobacteriacea. Except for the LPS from the non-enterobacteria *P. aeruginosa* and *V. cholerae*, the minimal LPS concentration to induce significant cytokine release, laid within one log-order for each cytokine ranging from 0.01-1 ng LPS/ml. The immune stimulatory activity of LPS from *P. aeruginosa* and *V. cholerae* was about a thousand-fold weaker and even at these high concentrations not the whole spectrum of cytokines was induced. Both LPS failed to induce IL-10, a finding fitting very well to the observation in the first part of the theses that a stronger stimulation of the monocytes is necessary to trigger IL-10 release. The reason for this decreased cytokine induction clearly can be found in the structure of the lipid A. In contrast to the sterical properties of LPS from enterobacteriacea, allowing an optimal receptor-ligand binding, the LPS from *P. aeruginosa,* and *V. cholerae* differ in the fatty acid chain length and the type of hydroxylated and nonhydroxylated fatty acids [79]. The LPS from *P. aeruginosa* contains only five fatty acids that are shorter in length (10 to 12 carbons) [150]. In our study, the *P. aeruginosa* LPS showed a TLR2 dependency in line with a previous report [157]. However, we could separate the TLR2-dependent lipoprotein fraction by phenol re-extraction. Obviously, for weak endotoxins, the immune stimulatory effects of contaminating agents are much more pronounced. In this light, studies employing crude *P. aeruginosa* LPS preparations with regard to TLR dependency should be reassessed. Because of *P. aeruginosa*'s ubiquitous occurrence and ability to grow in water at 4°C, they are a major source of contamination in purified water, especially of dialysis solutions or water prepared for injection purposes [170, 171]. They play an important role as opportunistic bacteria for patients with AIDS and for neutropenic patients, for infections of burn wounds, for lung infections in patients with nosocomial pneumonia or cystic fibrosis and acute ulcerative keratitis [172]. This observed weak endotoxic activity of *P. aeruginosa* might explain the fact that although *P. aeruginosa* shows a widespread occurrence in the environment, nearly all clinical cases are associated with immune compromised hosts and infection is rare in healthy people [178, 179]. Remarkably, while by LAL highly purified *P. aeruginosa* LPS was a thousand-fold less potent with regard to cytokine release from human whole blood, it was only half as potent as reference preparations from *S. abortus equi* or *E. coli*. A 10- to 100-fold weaker pyrogenicity in rabbits compared to other LPS like the ones from *E. coli* or *S. typhimurium* has also been reported [181]. Given the fact that the vast majority of LAL testing is carried out on water samples, this discrepancy for the foremost water contaminant is of critical importance. This is just another example that clearly indicates that the extent of activation of the Limulus cascade cannot be directly correlated with the pyrogenic potential for humans. This finding, where the endotoxic properties are overestimated by the LAL, contrasts the more frequent cases where endotoxic activities, even of synthetic Lipid A analogs, are not detected by LAL [81, 170, 182].

One kind of Gram- positive bacteria with a unique opportunistic intracellular infection cycle and immune evasion strategy is *L. monocytogenes* - the cause of food-borne listeriosis which has a fatality rate of up to 30%, particularly among infants, children and the elderly. The most frequent effects are meningitis and miscarriage or meningitis of the foetus or newborn. Although their incidence is relatively low, their severe and sometimes fatal health consequences make them among the most serious food-borne infections [109, 110]. Despite numerous publications about the regulation of virulence and immune responses towards *L. monocytogenes*, aspects of immune recognition by PRRs as well as structural differences between saprophytic (extracellular) and pathogenic (mostly intracellular) *Listeria* have hardly been examined. We show that 1.) Intracellular culture of *L. monocytogenes* leads to a significantly induced release of proinflammatory cytokines compared to cultivation under cell-free conditions in shaking flasks at room temperature. 2.) Temperature alone is capable of affecting the immune stimulatory potential of *Listeria*. 3.) *Listeria* express two variants of LTA and the decrease in the immune stimulatory potential is associated with a ten-fold increase of LTA2 expression. 4.) LTA2 is a much weaker inducer of cytokines and C4 turnover compared to LTA1. These changes in immunogenicity might serve to protect *L. monocytogenes* from the attacks of the host's innate immune system, for instance if they are set free from dying host cells. Additionally, the reduction of listericidal responses inside of activated macrophages could be of great relevance.

At the moment, it is speculative whether evolution has driven the expression of LTA2 for the benefit of immune evasion. Alternatively, this could be a secondary effect of a structural modification, which in first line, meets the requirements of cell division and cell wall homeostasis in the chemically and physically different environment of human host cell compartments. Since this alternative structure is much less expressed at low temperature and so far unique for the virulent *L. monocytogenes*, we may assume that for a saprophytic bacterium such modifications of LTA structure are of disadvantage, if constitutively expressed. A promising candidate for activating the switch in the LTA expression pattern was PrfA. It is the most prominent regulatory virulence factor in *Listeria* responsible for the regulation of a variety of genes involved in infection, intracellular growth and spread, also affecting cell wall components. Furthermore, analogous to LTA2 expression, its transcription is to a large degree temperature-dependent [113]. However, our findings with *prfA*-mutant *Listeria* strains denied any correlation. Since the syntheses pathways of listerial LTAs are fairly understood, it's difficult to identify the events leading to modifications. Noteworthy, another virulence factor reported recently, VirR, controls the modification of cell surface components and regulates the *dlt* operon which is involved in alanilation of LTA [200]. Although there is no difference in the degree of alanilation between LTA1 and LTA2, VirR would be the most obvious candidate to search further influences on LTA expression. Unfortunately, most intracellular pathogens with more or less comparable infection characteristics are not belonging to the Gram- positive bacteria; therefore, alternative study objects to investigate modulations of LTA expression are lacking. The adaptiveness uniquely displayed by the only human pathogenic *Listeria* strain is amazing, regarding the elaborately accomplished switch from the saprophytic to the pathogenic phenotype. Nevertheless, a single procedure modifying the indispensable cell wall molecule LTA seems to join the benefits of efficient intracellular growth and immune evasion.

This thesis considers some important factors determining the complexity of infectious disease. It shows that on host-side it is important to focus the investigations not only on genetic polymorphisms but to perform more global studies, while on the pathogen-side, small variations of PAMPs already have dramatic consequences. This, together with the increasing knowledge that is gained today about disease susceptibilities and treatment responses, to establish an overall, fundamental understanding of the mechanisms involved in virulence and immunity.

8 Summary

The courses of infections, ranging from subclinical to lethal diseases, are determined by numerous variables of the host defence and of the nature of the pathogen. Genetic polymorphisms, concerning both regulatory promoter and coding sequences, are increasingly linked with disease susceptibility and outcome. The first part of the study was attributed to a variant of one of the human pattern recognition receptors, which constitute key elements for the initiation of an innate immune response. The heterozygous TLR4 Asp(299)Gly polymorphism has been reported to be associated with inflammatory disorders, which were assumed to be a consequence of reduced proinflammatory responses due to the polymorphism. Unfortunately, cell-based assays are lacking in all studies. Our study, which aimed to prove deviations of inflammatory responses, revealed that after LPS stimulation of blood leukocytes only the expression of the anti-inflammatory cytokine IL-10 was reduced on mRNA and protein level, resulting in a proinflammatory phenotype. We could exclude that this effect is due to differences in the kinetics of IL-10 release, in the expression of total surface TLR4 or in LPS-binding to monocytes between subjects heterozygous for the Asp(299)Gly polymorphism or homozygous carriers of the wildtype allele. Furthermore, we could show that IL-10 induction in general requires stronger LPS-triggering than TNF and is more sensitive to LPS inhibitors. The lower number of responsive, wildtype TLR4 receptors on monocytes of heterozygotes may explain why only IL-10 expression is affected.

IL-10 dysfunction can result in excessive inflammation as shown for Crohn's disease, which is characterized by chronical or relapsing mucosal inflammation. We have investigated by incubations of human whole blood whether in Crohn's disease patients LPS-inducible TNF or IL-10 release is influenced by a heterozygous TLR4 polymorphism compared to patients with a homozygous TLR4 wildtype phenotype. Genotyping the patients and healthy controls for the *tlr4* gene, we could confirm the prevalence of the Asp(299)Gly polymorphism in the patients collective. Surprisingly, neither TNF nor IL-10 release was significantly different between both patient groups, and was furthermore comparable to cytokine release levels of healthy individuals. This indicates that probably at this stage of disease deviations in cytokine release occur only at the inflamed mucosa and cannot be detected by stimulations of leukocytes taken from the peripheral blood.

The focus of the second part of the study was the structural analysis of bacterial cell wall molecules, which are crucial for the initiation of immune defence and which also determine the extent of inflammatory reactions. Hereby we have examined non-classical cell wall molecules with endotoxic properties. We compared the immune stimulatory potency and TLR-dependency of eleven different LPS. While the majority of LPS induced cytokine release equipotentially, a thousand-fold more LPS from *Pseudomonas aeruginosa* or *Vibrio cholerae* was still less potent, though it potently induced ng quantities IL-8. All LPS tested, regardless of the microorganism, showed toll-like receptor (TLR)4-dependence, except for the LPS from *P. aeruginosa* and *V. cholerae,* which were both TLR4 and TLR2 dependent. However, re-purification of the commercial LPS preparation by phenol re-extraction led to a complete loss of the TLR2 dependency, indicating contaminations with lipoproteins. The Limulus Amebocyte Lysate Assay, often performed to exclude contaminations in purified water likely to originate from *P. aeruginosa,* resulted in an overestimation of pyrogenic burden of *P. aeruginosa* LPS by a factor 500 in the sample when compared with the biological activity of highly purified *P. aeruginosa* LPS in human whole blood.

Another aspect was *Listeria monocytogenes*, a food-borne, opportunistic intracellular pathogen. When human blood leukocytes were stimulated with *L. monocytogenes* that had either been grown in shaking flasks at room temperature (RT) or 37°C or intracellular in monocytic mass cultures, cytokine induction was strongest for *L. monocytogenes* grown at RT and weakest for intracellular grown bacteria. LTA extraction revealed that surprisingly all *Listeria*, independently from the type of culture condition, expressed two structurally different LTA. While the immune stimulatory potential of LTA1 was comparable to that of other LTA, LTA2 was significantly weaker in inducing cytokine release and even failed to induce complement activation. In line with our observation that intracellular grown *L. monocytogenes* possess a reduced proinflammatory potential, intracellular grown bacteria expressed LTA1 and LTA2 in a mass relation of 1.5:1, while grown cell-free at room temperature express a relation of 16:1, indicating that a shift in expression of structural variants of LTA might be an important mechanisms to hide from the innate immune system.

This thesis contributes to understand the consequences of molecular variations for host-pathogen interaction and immune defence. Especially the variations in LTA structure during intracellular adaptation of *Listeria* offers new targets for bactericidal therapies.

9 Abbreviations

A	adenosine
AIDS	aquired immune-deficiency syndrome
Asp	asparagine
ATCC	American Type Culture Collection
bp	base pair
CFU	colony forming unit
DC	dendritic cell
DNA	desoxyribonucleic acid
ds	double strand
DSMZ	Deutsche Sammlung für Mikroorganismen und Zellkulturen
DTT	dithiothreitol
EDTA	ethylendiamine tetra-acetate
ELISA	enzyme-linked immunosorbant assay
FACS	fluorescense-activated cell sorter
FCS	fetal calf serum
FPLC	fast protein liquid chromatography
FSC	forward scatter
G	guanidine
Gly	glycine
HIC	Hydrophobic Interaction Chromatography
IFN	interferon
Ile	isoleucine
Ig	immunoglobulin
IL	interleukin
Inl	internalin
LAL	limulus-amoebocyte-lysate
LALF	limulus anti-LPS factor
LLO	listeriolysin O
LPS	lipopolysaccharide
LTA	lipo-teichoic acid
MOI	multiplicity of infection

mRNA	messenger RNA
MS	mass spectrometry
NO	nitric oxide
NOD	nucleotide-binding oligomerization domain
OD	optical density
PAMP	pathogen associated molecular pattern
PBMC	peripheral blood mononuclear cells
PBS	phosphate buffered saline
PCR	polymerase chain reaction
Plc	phospholipase C
POD	horse-raddish-peroxidase
PRR	pattern recognition receptor
RFLP	restriction fragment length polymorphism
RNA	ribonucleic acid
rRNA	ribosomal RNA
rpm	rotations per minutes
RT	room temperature
RT-PCR	reverse transcription PCR
SDS	sodium dodecyl sulfat
SEM	standard error of the mean
Thr	threonine
TLR	toll-like rezeptor
TMB	3,3´5,5´-tetramethylbenzidine
TNF	tumor-necrosis-factor
Tris	Tris-(hydroxymethyl)-aminoethan
UV	ultra violet

10 References

1. Hoffmann, J.A., et al., *Phylogenetic perspectives in innate immunity.* Science, 1999. **284**(5418): p. 1313-8.
2. Petranyi, G.G., *The complexity of immune and alloimmune response.* Transpl Immunol, 2002. **10**(2-3): p. 91-100.
3. Akira, S. and H. Hemmi, *Recognition of pathogen-associated molecular patterns by TLR family.* Immunol Lett, 2003. **85**(2): p. 85-95.
4. Janeway, C.A., Jr. and R. Medzhitov, *Introduction: the role of innate immunity in the adaptive immune response.* Semin Immunol, 1998. **10**(5): p. 349-50.
5. Aderem, A. and R.J. Ulevitch, *Toll-like receptors in the induction of the innate immune response.* Nature, 2000. **406**(6797): p. 782-7.
6. Michalek, S.M., et al., *The primary role of lymphoreticular cells in the mediation of host responses to bacterial endotoxim.* J Infect Dis, 1980. **141**(1): p. 55-63.
7. Freudenberg, M.A., D. Keppler, and C. Galanos, *Requirement for lipopolysaccharide-responsive macrophages in galactosamine-induced sensitization to endotoxin.* Infect Immun, 1986. **51**(3): p. 891-5.
8. Akira, S., K. Takeda, and T. Kaisho, *Toll-like receptors: critical proteins linking innate and acquired immunity.* Nat Immunol, 2001. **2**(8): p. 675-80.
9. Tomlinson, S., *Complement defense mechanisms.* Curr Opin Immunol, 1993. **5**(1): p. 83-9.
10. Marsh, C.B. and M.D. Wewers, *The pathogenesis of sepsis. Factors that modulate the response to gram-negative bacterial infection.* Clin Chest Med, 1996. **17**(2): p. 183-97.
11. Offner, P.J., E.E. Moore, and W.L. Biffl, *Male gender is a risk factor for major infections after surgery.* Arch Surg, 1999. **134**(9): p. 935-8; discussion 938-40.
12. Ono, S., et al., *Sex differences in cytokine production and surface antigen expression of peripheral blood mononuclear cells after surgery.* Am J Surg, 2005. **190**(3): p. 439-44.
13. Aulock, S.V., et al., *Gender difference in cytokine secretion on immune stimulation with LPS and LTA.* J Interferon Cytokine Res, 2006. **26**(12): p. 887-92.
14. Gathof, B.S., S.M. Picker, and J. Rojo, *Epidemiology, etiology and diagnosis of venous thrombosis.* Eur J Med Res, 2004. **9**(3): p. 95-103.

15. Calandra, T., P.Y. Bochud, and D. Heumann, *Cytokines in septic shock.* Curr Clin Top Infect Dis, 2002. **22**: p. 1-23.
16. Opal, S.M. and T. Gluck, *Endotoxin as a drug target.* Crit Care Med, 2003. **31**(1 Suppl): p. S57-64.
17. Taveira da Silva, A.M., et al., *Brief report: shock and multiple-organ dysfunction after self-administration of Salmonella endotoxin.* N Engl J Med, 1993. **328**(20): p. 1457-60.
18. Bochud, P.Y., et al., *Community-acquired pneumonia. A prospective outpatient study.* Medicine (Baltimore), 2001. **80**(2): p. 75-87.
19. Tracey, K.J., et al., *Anti-cachectin/TNF monoclonal antibodies prevent septic shock during lethal bacteraemia.* Nature, 1987. **330**(6149): p. 662-4.
20. Beutler, B. and A. Cerami, *Cachectin: more than a tumor necrosis factor.* N Engl J Med, 1987. **316**(7): p. 379-85.
21. Mannel, D.N., R.N. Moore, and S.E. Mergenhagen, *Macrophages as a source of tumoricidal activity (tumor-necrotizing factor).* Infect Immun, 1980. **30**(2): p. 523-30.
22. Conti, P., et al., *IL-10, an inflammatory/inhibitory cytokine, but not always.* Immunol Lett, 2003. **86**(2): p. 123-9.
23. Kontoyiannis, D., et al., *Impaired on/off regulation of TNF biosynthesis in mice lacking TNF AU-rich elements: implications for joint and gut-associated immunopathologies.* Immunity, 1999. **10**(3): p. 387-98.
24. Agnese, D.M., et al., *Human toll-like receptor 4 mutations but not CD14 polymorphisms are associated with an increased risk of gram-negative infections.* J Infect Dis, 2002. **186**(10): p. 1522-5.
25. Schottelius, A.J., et al., *Interleukin-10 signaling blocks inhibitor of kappaB kinase activity and nuclear factor kappaB DNA binding.* J Biol Chem, 1999. **274**(45): p. 31868-74.
26. Hart, P.H., et al., *Regulation of surface and soluble TNF receptor expression on human monocytes and synovial fluid macrophages by IL-4 and IL-10.* J Immunol, 1996. **157**(8): p. 3672-80.
27. Dickensheets, H.L., et al., *Interleukin-10 upregulates tumor necrosis factor receptor type-II (p75) gene expression in endotoxin-stimulated human monocytes.* Blood, 1997. **90**(10): p. 4162-71.
28. Jenkins, J.K., M. Malyak, and W.P. Arend, *The effects of interleukin-10 on interleukin-1 receptor antagonist and interleukin-1 beta production in human monocytes and neutrophils.* Lymphokine Cytokine Res, 1994. **13**(1): p. 47-54.

29. Gerard, C., et al., *Interleukin 10 reduces the release of tumor necrosis factor and prevents lethality in experimental endotoxemia.* J Exp Med, 1993. **177**(2): p. 547-50.
30. Marchant, A., et al., *Interleukin-10 controls interferon-gamma and tumor necrosis factor production during experimental endotoxemia.* Eur J Immunol, 1994. **24**(5): p. 1167-71.
31. Kuhn, R., et al., *Interleukin-10-deficient mice develop chronic enterocolitis.* Cell, 1993. **75**(2): p. 263-74.
32. van Dissel, J.T., et al., *Anti-inflammatory cytokine profile and mortality in febrile patients.* Lancet, 1998. **351**(9107): p. 950-3.
33. Adib-Conquy, M.A., C., et al., *NF-kappaB expression in mononuclear cells of patients with sepsis resembles that observed in lipopolysaccharide tolerance.* Am J Respir Crit Care Med, 2000. **162**(5): p. 1877-83.
34. Adib-Conquy, M., et al., *Long-term-impaired expression of nuclear factor-kappa B and I kappa B alpha in peripheral blood mononuclear cells of trauma patients.* J Leukoc Biol, 2001. **70**(1): p. 30-8.
35. Gibson, A.W., et al., *Novel single nucleotide polymorphisms in the distal IL-10 promoter affect IL-10 production and enhance the risk of systemic lupus erythematosus.* J Immunol, 2001. **166**(6): p. 3915-22.
36. Tagore, A., et al., *Interleukin-10 (IL-10) genotypes in inflammatory bowel disease.* Tissue Antigens, 1999. **54**(4): p. 386-90.
37. Castro-Santos, P., et al., *TNFalpha and IL-10 gene polymorphisms in inflammatory bowel disease. Association of -1082 AA low producer IL-10 genotype with steroid dependency.* Am J Gastroenterol, 2006. **101**(5): p. 1039-47.
38. van Hogezand, R.A. and H.W. Verspaget, *New therapies for inflammatory bowel disease: an update on chimeric anti-TNF alpha antibodies and IL-10 therapy.* Scand J Gastroenterol Suppl, 1997. **223**: p. 105-7.
39. Opal, S.M., J.C. Wherry, and P. Grint, *Interleukin-10: potential benefits and possible risks in clinical infectious diseases.* Clin Infect Dis, 1998. **27**(6): p. 1497-507.
40. Lien, E., et al., *Toll-like receptor 4 imparts ligand-specific recognition of bacterial lipopolysaccharide.* J Clin Invest, 2000. **105**(4): p. 497-504.
41. Zhang, D., et al., *A toll-like receptor that prevents infection by uropathogenic bacteria.* Science, 2004. **303**(5663): p. 1522-6.
42. Hashimoto, C., K.L. Hudson, and K.V. Anderson, *The Toll gene of Drosophila, required for dorsal-ventral embryonic polarity, appears to encode a transmembrane protein.* Cell, 1988. **52**(2): p. 269-79.

43. Ligoxygakis, P., et al., *Activation of Drosophila Toll during fungal infection by a blood serine protease.* Science, 2002. **297**(5578): p. 114-6.
44. Takeda, K., T. Kaisho, and S. Akira, *Toll-like receptors.* Annu Rev Immunol, 2003. **21**: p. 335-76.
45. Wyllie, D.H., et al., *Evidence for an accessory protein function for Toll-like receptor 1 in anti-bacterial responses.* J Immunol, 2000. **165**(12): p. 7125-32.
46. Takeuchi, O., et al., *Cutting edge: role of Toll-like receptor 1 in mediating immune response to microbial lipoproteins.* J Immunol, 2002. **169**(1): p. 10-4.
47. Schwandner, R., et al., *Peptidoglycan- and lipoteichoic acid-induced cell activation is mediated by toll-like receptor 2.* J Biol Chem, 1999. **274**(25): p. 17406-9.
48. Alexopoulou, L., et al., *Recognition of double-stranded RNA and activation of NF-kappaB by Toll-like receptor 3.* Nature, 2001. **413**(6857): p. 732-8.
49. Hoshino, K., et al., *Cutting edge: Toll-like receptor 4 (TLR4)-deficient mice are hyporesponsive to lipopolysaccharide: evidence for TLR4 as the Lps gene product.* J Immunol, 1999. **162**(7): p. 3749-52.
50. Poltorak, A., et al., *Defective LPS signaling in C3H/HeJ and C57BL/10ScCr mice: mutations in Tlr4 gene.* Science, 1998. **282**(5396): p. 2085-8.
51. Hayashi, F., et al., *The innate immune response to bacterial flagellin is mediated by Toll-like receptor 5.* Nature, 2001. **410**(6832): p. 1099-103.
52. Takeuchi, O., et al., *TLR6: A novel member of an expanding toll-like receptor family.* Gene, 1999. **231**(1-2): p. 59-65.
53. Heil, F., et al., *Species-specific recognition of single-stranded RNA via toll-like receptor 7 and 8.* Science, 2004. **303**(5663): p. 1526-9.
54. Hemmi, H., et al., *A Toll-like receptor recognizes bacterial DNA.* Nature, 2000. **408**(6813): p. 740-5.
55. Chuang, T. and R.J. Ulevitch, *Identification of hTLR10: a novel human Toll-like receptor preferentially expressed in immune cells.* Biochim Biophys Acta, 2001. **1518**(1-2): p. 157-61.
56. Medzhitov, R., P. Preston-Hurlburt, and C.A. Janeway, Jr., *A human homologue of the Drosophila Toll protein signals activation of adaptive immunity.* Nature, 1997. **388**(6640): p. 394-7.
57. Poltorak, A., et al., *Genetic and physical mapping of the Lps locus: identification of the toll-4 receptor as a candidate gene in the critical region.* Blood Cells Mol Dis, 1998. **24**(3): p. 340-55.

58. Liu, Y.J., *Dendritic cell subsets and lineages, and their functions in innate and adaptive immunity.* Cell, 2001. **106**(3): p. 259-62.
59. Visintin, A., et al., *Regulation of Toll-like receptors in human monocytes and dendritic cells.* J Immunol, 2001. **166**(1): p. 249-55.
60. Malaviya, R. and S.N. Abraham, *Mast cell modulation of immune responses to bacteria.* Immunol Rev, 2001. **179**: p. 16-24.
61. Abreu, M.T., et al., *Decreased expression of Toll-like receptor-4 and MD-2 correlates with intestinal epithelial cell protection against dysregulated proinflammatory gene expression in response to bacterial lipopolysaccharide.* J Immunol, 2001. **167**(3): p. 1609-16.
62. Schroder, N.W. and R.R. Schumann, *Single nucleotide polymorphisms of Toll-like receptors and susceptibility to infectious disease.* Lancet Infect Dis, 2005. **5**(3): p. 156-64.
63. Santamaria, P., et al., *Involvement of class II MHC molecules in the LPS-induction of IL-1/TNF secretions by human monocytes. Quantitative differences at the polymorphic level.* J Immunol, 1989. **143**(3): p. 913-22.
64. Arbour, N.C., et al., *TLR4 mutations are associated with endotoxin hyporesponsiveness in humans.* Nat Genet, 2000. **25**(2): p. 187-91.
65. Smirnova, I., et al., *Excess of rare amino acid polymorphisms in the Toll-like receptor 4 in humans.* Genetics, 2001. **158**(4): p. 1657-64.
66. Franchimont, D., et al., *Deficient host-bacteria interactions in inflammatory bowel disease? The toll-like receptor (TLR)-4 Asp299gly polymorphism is associated with Crohn's disease and ulcerative colitis.* Gut, 2004. **53**(7): p. 987-92.
67. Torok, H., et al., *Polymorphisms of the lipopolysaccharide-signaling complex in inflammatory bowel disease: association of a mutation in the Toll-like receptor 4 gene with ulcerative colitis.* Clinical Immunology, 2004. **112**(1): p. 85-91.
68. Kleemola, M., et al., *Epidemics of pneumonia caused by TWAR, a new Chlamydia organism, in military trainees in Finland.* J Infect Dis, 1988. **157**(2): p. 230-6.
69. Kreger, B.E., et al., *Gram-negative bacteremia. III. Reassessment of etiology, epidemiology and ecology in 612 patients.* Am J Med, 1980. **68**(3): p. 332-43.
70. Bryant, R.E., et al., *Factors affecting mortality of gram-negative rod bacteremia.* Arch Intern Med, 1971. **127**(1): p. 120-8.
71. Bryan, C.S., K.L. Reynolds, and E.R. Brenner, *Analysis of 1,186 episodes of gram-negative bacteremia in non-university hospitals: the effects of antimicrobial therapy.* Rev Infect Dis, 1983. **5**(4): p. 629-38.

72. Rakoff-Nahoum, S. and R. Medzhitov, *Role of the innate immune system and host-commensal mutualism.* Curr Top Microbiol Immunol, 2006. **308**: p. 1-18.
73. Flamez, C., et al., *Two-component system regulon plasticity in bacteria: a concept emerging from phenotypic analysis of Yersinia pseudotuberculosis response regulator mutants.* Adv Exp Med Biol, 2007. **603**: p. 145-55.
74. Morath, S., S. von Aulock, and T. Hartung, *Structure/function relationships of lipoteichoic acids.* J Endotoxin Res, 2005. **11**(6): p. 348-56.
75. sMorath, S., A. Geyer, and T. Hartung, *Structure-function relationship of cytokine induction by lipoteichoic acid from Staphylococcus aureus.* J Exp Med, 2001. **193**(3): p. 393-7.
76. Demchick, P. and A.L. Koch, *The permeability of the wall fabric of Escherichia coli and Bacillus subtilis.* J Bacteriol, 1996. **178**(3): p. 768-73.
77. Rietschel, E.T., et al., *Bacterial endotoxin: molecular relationships of structure to activity and function.* FASEB J., 1994. **8**: p. 217-225.
78. Rietschel, E.T., et al., *Lipid A, the endotoxic center of bacterial lipopolysaccharides: relation of chemical structure to biological activity.* Prog Clin Biol Res, 1987. **231**: p. 25-53.
79. Rietschel, E.T., et al., *Structure and Function of the Lipid A component of lipopolysaccharides.* Handbook of Endotoxins, 1984. **1**(Chemistry of endotoxin): p. 187-220.
80. Galanos, C., et al., *Synthetic and natural Escherichia coli free lipid A express identical endotoxic activities.* Eur J Biochem, 1985. **148**(1): p. 1-5.
81. Tanamoto, K., et al., *Biological activities of synthetic lipid A analogs: pyrogenicity, lethal toxicity, anticomplement activity, and induction of gelation of Limulus amoebocyte lysate.* Infect Immun, 1984. **44**(2): p. 421-6.
82. Netea, M.G., et al., *Does the shape of lipid A determine the interaction of LPS with Toll-like receptors?* Trends Immunol, 2002. **23**(3): p. 135-9.
83. Caroff, M., et al., *Structural and functional analyses of bacterial lipopolysaccharides.* Microbes Infect, 2002. **4**(9): p. 915-26.
84. Loppnow, H., et al., *Cytokine induction by lipopolysaccharide (LPS) corresponds to lethal toxicity and is inhibited by nontoxic Rhodobacter capsulatus LPS.* Infect Immun, 1990. **58**(11): p. 3743-50.
85. Werts, C., et al., *Leptospiral lipopolysaccharide activates cells through a TLR2-dependent mechanism.* Nat Immunol, 2001. **2**(4): p. 346-52.

86. Vogel, S., M.J. Hirschfeld, and P.Y. Perera, *Signal integration in lipopolysaccharide (LPS)-stimulated murine macrophages.* J Endotoxin Res, 2001. **7**(3): p. 237-41.
87. Ernst, R.K., et al., *Specific lipopolysaccharide found in cystic fibrosis airway Pseudomonas aeruginosa.* Science, 1999. **286**(5444): p. 1561-5.
88. Fischetti, V. and R. Novick, et al., eds. *Gram-positive pathogens.* 2 ed. 2006, ASM Press, Washington, DC, USA.
89. Weidenmaier, C., et al., *Role of teichoic acids in Staphylococcus aureus nasal colonization, a major risk factor in nosocomial infections.* Nat Med, 2004. **10**(3): p. 243-5.
90. Weidenmaier, C., et al., *Lack of wall teichoic acids in Staphylococcus aureus leads to reduced interactions with endothelial cells and to attenuated virulence in a rabbit model of endocarditis.* J Infect Dis, 2005. **191**(10): p. 1771-7.
91. Grundling, A. and O. Schneewind, *Genes required for glycolipid synthesis and lipoteichoic acid anchoring in Staphylococcus aureus.* J Bacteriol, 2007. **189**(6): p. 2521-30.
92. Fedtke, I., et al., *A Staphylococcus aureus ypfP mutant with strongly reduced lipoteichoic acid (LTA) content: LTA governs bacterial surface properties and autolysin activity.* Mol Microbiol, 2007. **65**(4): p. 1078-91.
93. Hughes, A.H., I.C. Hancock, and J. Baddiley, *The function of teichoic acids in cation control in bacterial membranes.* Biochem J, 1973. **132**(1): p. 83-93.
94. Peschel, A., et al., *Inactivation of the dlt operon in Staphylococcus aureus confers sensitivity to defensins, protegrins, and other antimicrobial peptides.* J Biol Chem, 1999. **274**(13): p. 8405-10.
95. Cabanes, D., et al., *Surface proteins and the pathogenic potential of Listeria monocytogenes.* Trends Microbiol, 2002. **10**(5): p. 238-45.
96. Jonquieres, R., et al., *Interaction between the protein InlB of Listeria monocytogenes and lipoteichoic acid: a novel mechanism of protein association at the surface of gram-positive bacteria.* Mol Microbiol, 1999. **34**(5): p. 902-14.
97. Morath, S., et al., *Structural decomposition and heterogeneity of commercial lipoteichoic Acid preparations.* Infect Immun, 2002. **70**(2): p. 938-44.
98. Lynch, N.J., et al., *L-ficolin specifically binds to lipoteichoic acid, a cell wall constituent of Gram-positive bacteria, and activates the lectin pathway of complement.* J Immunol, 2004. **172**(2): p. 1198-202.

99. Morath, S., A. Geyer, and T. Hartung, *Structure-function relationship of cytokine induction by lipoteichoic acid from Staphylococcus aureus.* J Exp Med, 2001. **193**(3): p. 393-7.
100. Draing, C., et al., *Comparison of lipoteichoic acid from different serotypes of Streptococcus pneumoniae.* J Biol Chem, 2006. **281**(45): p. 33849-59.
101. Grangette, C., et al., *Enhanced antiinflammatory capacity of a Lactobacillus plantarum mutant synthesizing modified teichoic acids.* Proc Natl Acad Sci U S A, 2005. **102**(29): p. 10321-6.
102. Hermann, C., J Endotoxin Res, 2008.
103. Deininger, S., et al., *Definition of Structural Prerequisites for Lipoteichoic Acid-inducible Cytokine Induction by Synthetic Derivates.* J Immunol, 2003. **170**: p. in press.
104. Girardin, S.E., et al., *Nod2 is a general sensor of peptidoglycan through muramyl dipeptide (MDP) detection.* J Biol Chem, 2003. **278**(11): p. 8869-72.
105. Girardin, S.E., et al., *Nod1 detects a unique muropeptide from gram-negative bacterial peptidoglycan.* Science, 2003. **300**(5625): p. 1584-7.
106. Inohara, et al., *NOD-LRR proteins: role in host-microbial interactions and inflammatory disease.* Annu Rev Biochem, 2005. **74**: p. 355-83.
107. Ganten, D., Ruckpaul, K., ed. *immunsystem und Infektiologie.* 1999, Springer.
108. Farber, J.M. and P.I. Peterkin, *Listeria monocytogenes, a food-borne pathogen.* Microbiol Rev, 1991. **55**(3): p. 476-511.
109. Vazquez-Boland, J.A., et al., *Listeria pathogenesis and molecular virulence determinants.* Clin Microbiol Rev, 2001. **14**(3): p. 584-640.
110. Ramaswamy, V., et al., *Listeria--review of epidemiology and pathogenesis.* J Microbiol Immunol Infect, 2007. **40**(1): p. 4-13.
111. Jackson, L.A., et al., *Isolation of Chlamydia pneumoniae from a carotid endarterectomy specimen.* J Infect Dis, 1997. **176**(1): p. 292-5.
112. Farber, J.M., et al., *Neonatal listeriosis due to cross-infection confirmed by isoenzyme typing and DNA fingerprinting.* J Infect Dis, 1991. **163**(4): p. 927-8.
113. Scortti, M., et al., *The PrfA virulence regulon.* Microbes Infect, 2007. **9**(10): p. 1196-207.
114. Port, G.C. and N.E. Freitag, *Identification of novel Listeria monocytogenes secreted virulence factors following mutational activation of the central virulence regulator, PrfA.* Infect Immun, 2007. **75**(12): p. 5886-97.

115. Calvano, J.E., et al., *Response to systemic endotoxemia among humans bearing polymorphisms of the Toll-like receptor 4 (hTLR4)*. Clin Immunol, 2006. **121**(2): p. 186-90.
116. Schippers, E.F., et al., *TNF-alpha promoter, Nod2 and toll-like receptor-4 polymorphisms and the in vivo and ex vivo response to endotoxin*. Cytokine, 2004. **26**(1): p. 16-24.
117. Hermann, C., *Variability of host-pathogen interaction*. J Endotoxin Res, 2007. **13**(4): p. 199-218.
118. Erridge, C., J. Stewart, and I.R. Poxton, *Monocytes heterozygous for the Asp299Gly and Thr399Ile mutations in the Toll-like receptor 4 gene show no deficit in lipopolysaccharide signalling*. J Exp Med, 2003. **197**(12): p. 1787-91.
119. von Aulock, S., et al., *Heterozygous toll-like receptor 4 polymorphism does not influence lipopolysaccharide-induced cytokine release in human whole blood*. J Infect Dis, 2003. **188**(6): p. 938-43.
120. Moore, K.W., et al., *Interleukin-10 and the interleukin-10 receptor*. Annu Rev Immunol, 2001. **19**: p. 683-765.
121. Heesen, M., et al., *Rapid and reliable genotyping for the Toll-like receptor 4 A896G polymorphism using fluorescence-labeled hybridization probes in a real-time polymerase chain reaction assay*. Clin Chim Acta, 2003. **333**(1): p. 47-9.
122. Rallabhandi, P., et al., *Analysis of TLR4 polymorphic variants: new insights into TLR4/MD-2/CD14 stoichiometry, structure, and signaling*. J Immunol, 2006. **177**(1): p. 322-32.
123. Hoebe, K. and B. Beutler, *LPS, dsRNA and the interferon bridge to adaptive immune responses: Trif, Tram, and other TIR adaptor proteins*. J Endotoxin Res, 2004. **10**(2): p. 130-6.
124. Ma, W., et al., *The p38 mitogen-activated kinase pathway regulates the human interleukin-10 promoter via the activation of Sp1 transcription factor in lipopolysaccharide-stimulated human macrophages*. J Biol Chem, 2001. **276**(17): p. 13664-74.
125. Adib-Conquy, M., et al., *Toll-like receptor-mediated tumor necrosis factor and interleukin-10 production differ during systemic inflammation*. Am J Respir Crit Care Med, 2003. **168**(2): p. 158-64.
126. Hacker, H., et al., *Specificity in Toll-like receptor signalling through distinct effector functions of TRAF3 and TRAF6*. Nature, 2006. **439**(7073): p. 204-7.

127. Williams, J.A., C.H. Pontzer, and E. Shacter, *Regulation of macrophage interleukin-6 (IL-6) and IL-10 expression by prostaglandin E2: the role of p38 mitogen-activated protein kinase.* J Interferon Cytokine Res, 2000. **20**(3): p. 291-8.

128. Latz, E., et al., *Lipopolysaccharide rapidly traffics to and from the Golgi apparatus with the toll-like receptor 4-MD-2-CD14 complex in a process that is distinct from the initiation of signal transduction.* J Biol Chem, 2002. **277**(49): p. 47834-43.

129. Balish, E. and T. Warner, *Enterococcus faecalis induces inflammatory bowel disease in interleukin-10 knockout mice.* Am J Pathol, 2002. **160**(6): p. 2253-7.

130. Kim, S.C., et al., *Variable phenotypes of enterocolitis in interleukin 10-deficient mice monoassociated with two different commensal bacteria.* Gastroenterology, 2005. **128**(4): p. 891-906.

131. Rund, S., et al., *Structural analysis of the lipopolysaccharide from Chlamydia trachomatis serotype L2.* J Biol Chem, 1999. **274**(24): p. 16819-24.

132. Brand, S., et al., *The role of Toll-like receptor 4 Asp299Gly and Thr399Ile polymorphisms and CARD15/NOD2 mutations in the susceptibility and phenotype of Crohn's disease.* Inflamm Bowel Dis, 2005. **11**(7): p. 645-52.

133. Dehus, O., Bunk, S., von Aulock, S., Hermann, C., *IL-10 release requires stronger toll-like receptor 4-triggering than TNF- a possible explanation for the selective effects of heterozygous TLR4 polymorphism Asp(299)Gly on IL-10 release.* Immunbiology, 2008.

134. Schafmayer, C., et al., *Investigation of the Lith1 candidate genes ABCB11 and LXRA in human gallstone disease.* Hepatology, 2006. **44**(3): p. 650-7.

135. Hume, G.E., et al., *Novel NOD2 haplotype strengthens the association between TLR4 Asp299gly and Crohn's disease in an Australian population.* Inflamm Bowel Dis, 2008.

136. Browning, B.L., et al., *Has toll-like receptor 4 been prematurely dismissed as an inflammatory bowel disease gene? Association study combined with meta-analysis shows strong evidence for association.* Am J Gastroenterol, 2007. **102**(11): p. 2504-12.

137. Braat, H., M.P. Peppelenbosch, and D.W. Hommes, *Interleukin-10-based therapy for inflammatory bowel disease.* Expert Opin Biol Ther, 2003. **3**(5): p. 725-31.

138. van der Linde, K., et al., *A Gly15Arg mutation in the interleukin-10 gene reduces secretion of interleukin-10 in Crohn disease.* Scand J Gastroenterol, 2003. **38**(6): p. 611-7.

139. Rietschel, E.T. and H. Brade, *Bacterial endotoxins.* Sci Am, 1992. **267**(2): p. 54-61.

140. Vogel, S.N. and M.M. Hogan, *The role of cytokines in endotoxin-mediated host responses.* Immunopharmacology-the role of cells and cytokines in immunity and inflammation, 1990: p. 238-258.

141. Morrison, D.C. and J.L. Ryan, *Endotoxins and disease mechanisms.* Annu Rev Med, 1987. **38**: p. 417-32.

142. Raetz, C.R., *Biochemistry of endotoxins.* Annu Rev Biochem, 1990. **59**: p. 129-70.

143. Raetz, C.R., *Bacterial lipopolysaccharides: a remarkable family of bioactive macroamphiphiles.* Neidhart, F.C. : Escherichia coli and Salmonella: Cellular and molecular biology, 1996.

144. Loppnow, H., et al., *Induction of human interleukin 1 by bacterial and synthetic lipid A.* Eur J Immunol, 1986. **16**(10): p. 1263-7.

145. Takeda, K. and S. Akira, *Toll-like receptors in innate immunity.* Int Immunol, 2005. **17**(1): p. 1-14.

146. Watson, J. and R. Riblet, *Genetic control of responses to bacterial lipopolysaccharides in mice. I. Evidence for a single gene that influences mitogenic and immunogenic respones to lipopolysaccharides.* J Exp Med, 1974. **140**(5): p. 1147-61.

147. Kirschning, C.J., et al., *Human toll-like receptor 2 confers responsiveness to bacterial lipopolysaccharide.* J Exp Med, 1998. **188**(11): p. 2091-7.

148. Yang, R.B., et al., *Toll-like receptor-2 mediates lipopolysaccharide-induced cellular signalling.* Nature, 1998. **395**(6699): p. 284-8.

149. Hirschfeld, M., et al., *Cutting edge: repurification of lipopolysaccharide eliminates signaling through both human and murine toll-like receptor 2.* J Immunol, 2000. **165**(2): p. 618-22.

150. Erridge, C., E. Bennett-Guerrero, and I.R. Poxton, *Structure and function of lipopolysaccharides.* Microbes Infect, 2002. **4**(8): p. 837-51.

151. Lorenz, E., et al., *Toll-like receptor 4 (TLR4)-deficient murine macrophage cell line as an in vitro assay system to show TLR4-independent signaling of Bacteroides fragilis lipopolysaccharide.* Infect Immun, 2002. **70**(9): p. 4892-6.

152. Hirschfeld, M., et al., *Signaling by toll-like receptor 2 and 4 agonists results in differential gene expression in murine macrophages.* Infect Immun, 2001. **69**(3): p. 1477-82.

153. Hartung, T., *Die Erfassung von Pyrogenen in einem humanen Vollblutmodell.* Altex, 1995: p. 70-75.

154. Hartung, T., et al., *Effect of granulocyte colony-stimulating factor treatment on ex vivo blood cytokine response in human volunteers.* Blood, 1995. **85**(9): p. 2482-9.
155. Zughaier, S.M., H.C. Ryley, and S.K. Jackson, *Lipopolysaccharide (LPS) from Burkholderia cepacia is more active than LPS from Pseudomonas aeruginosa and Stenotrophomonas maltophilia in stimulating tumor necrosis factor alpha from human monocytes.* Infect Immun, 1999. **67**(3): p. 1505-7.
156. Mathiak, G., et al., *Lipopolysaccharides from different bacterial sources elicit disparate cytokine responses in whole blood assays.* Int J Mol Med, 2003. **11**(1): p. 41-4.
157. Erridge, C., et al., *Lipopolysaccharides of Bacteroides fragilis, Chlamydia trachomatis and Pseudomonas aeruginosa signal via toll-like receptor 2.* J Med Microbiol, 2004. **53**(Pt 8): p. 735-40.
158. Broady, K.W., E.T. Rietschel, and O. Luderitz, *The chemical structure of the lipid A component of lipopolysaccharides from Vibrio cholerae.* Eur J Biochem, 1981. **115**(3): p. 463-8.
159. Takayama, K. and N. Qureshi, *Chemical structure of lipid A.* Bacterial Endotoxic Lipopolysaccharides, 1992. **1**: p. 43-60.
160. Qureshi, N., K. Takayama, and R. Kurtz, *Diphosphoryl lipid A obtained from the nontoxic lipopolysaccharide of Rhodopseudomonas sphaeroides is an endotoxin antagonist in mice.* Infect Immun, 1991. **59**(1): p. 441-4.
161. Golenbock, D.T., et al., *Lipid A-like molecules that antagonize the effects of endotoxins on human monocytes.* J Biol Chem, 1991. **266**(29): p. 19490-8.
162. Schromm, A.B., et al., *Biological activities of lipopolysaccharides are determined by the shape of their lipid A portion.* Eur J Biochem, 2000. **267**(7): p. 2008-13.
163. Soong, G., et al., *TLR2 is mobilized into an apical lipid raft receptor complex to signal infection in airway epithelial cells.* J Clin Invest, 2004. **113**(10): p. 1482-9.
164. Adamo, R., et al., *Pseudomonas aeruginosa flagella activate airway epithelial cells through asialoGM1 and toll-like receptor 2 as well as toll-like receptor 5.* Am J Respir Cell Mol Biol, 2004. **30**(5): p. 627-34.
165. Sadikot, R.T., et al., *p47phox deficiency impairs NF-kappa B activation and host defense in Pseudomonas pneumonia.* J Immunol, 2004. **172**(3): p. 1801-8.
166. Faure, K., et al., *TLR4 signaling is essential for survival in acute lung injury induced by virulent Pseudomonas aeruginosa secreting type III secretory toxins.* Respir Res, 2004. **5**(1): p. 1.

167. Hajjar, A.M., et al., *Human Toll-like receptor 4 recognizes host-specific LPS modifications.* Nat Immunol, 2002. **3**(4): p. 354-9.
168. Green, S.K., et al., *Agricultural plants and soil as a reservoir for Pseudomonas aeruginosa.* Appl Microbiol, 1974. **28**(6): p. 987-91.
169. Williams, P.A. and M.J. Worsey, *Ubiquity of plasmids in coding for toluene and xylene metabolism in soil bacteria: evidence for the existence of new TOL plasmids.* J Bacteriol, 1976. **125**(3): p. 818-28.
170. Cappelli, G., et al., *Removal of limulus reactivity and cytokine-inducing capacity from bicarbonate dialysis fluids by ultrafiltration.* Nephrol Dial Transplant, 1993. **8**(10): p. 1133-9.
171. Sundaram, S., et al., *Transmembrane passage of cytokine-inducing bacterial products across new and reprocessed polysulfone dialyzers.* J Am Soc Nephrol, 1996. **7**(10): p. 2183-91.
172. Lyczak, J.B., C.L. Cannon, and G.B. Pier, *Establishment of Pseudomonas aeruginosa infection: lessons from a versatile opportunist.* Microbes Infect, 2000. **2**(9): p. 1051-60.
173. Ohman, D.E., R.P. Burns, and B.H. Iglewski, *Corneal infections in mice with toxin A and elastase mutants of Pseudomonas aeruginosa.* J Infect Dis, 1980. **142**(4): p. 547-55.
174. Woods, D.E., et al., *Contribution of toxin A and elastase to virulence of Pseudomonas aeruginosa in chronic lung infections of rats.* Infect. Immun., 1982. **36**: p. 1223-1228.
175. Buret, A. and A.W. Cripps, *The immunoevasive activities of Pseudomonas aeruginosa. Relevance for cystic fibrosis.* Am Rev Respir Dis, 1993. **148**(3): p. 793-805.
176. Cripps, A.W., et al., *Pulmonary immunity to Pseudomonas aeruginosa.* Immunol Cell Biol, 1995. **73**(5): p. 418-24.
177. Lin, T.J., et al., *Pseudomonas aeruginosa activates human mast cells to induce neutrophil transendothelial migration via mast cell-derived IL-1 alpha and beta.* J Immunol, 2002. **169**(8): p. 4522-30.
178. Franzetti, F., et al., *Pseudomonas infections in patients with AIDS and AIDS-related complex.* J Intern Med, 1992. **231**(4): p. 437-43.
179. Kielhofner, M., et al., *Life-threatening Pseudomonas aeruginosa infections in patients with human immunodeficiency virus infection.* Clin Infect Dis, 1992. **14**(2): p. 403-11.

180. Giamarellos-Bourboulis, E.J., et al., *Stimulation of innate immunity by susceptible and multidrug-resistant Pseudomonas aeruginosa: an in vitro and in vivo study.* Clin Exp Immunol, 2004. **135**(2): p. 240-6.
181. Tanamoto, K. and J.Y. Homma, *Essential regions of the lipopolysaccharide of Pseudomonas aeruginosa responsible for pyrogenicity and activation of the proclotting enzyme of horseshoe crabs. Comparison with antitumor, interferon-inducing and adjuvant activities.* J Biochem (Tokyo), 1982. **91**(3): p. 741-6.
182. Laude-Sharp, M., et al., *Dissociation between the interleukin 1-inducing capacity and Limulus reactivity of lipopolysaccharides from gram-negative bacteria.* Cytokine, 1990. **2**(4): p. 253-8.
183. Portnoy, D.A., V. Auerbuch, and I.J. Glomski, *The cell biology of Listeria monocytogenes infection: the intersection of bacterial pathogenesis and cell-mediated immunity.* J Cell Biol, 2002. **158**(3): p. 409-14.
184. Kreft, J., et al., *Pathogenicity islands and other virulence elements in Listeria.* Curr Top Microbiol Immunol, 2002. **264**(2): p. 109-25.
185. Wong, K.K. and N.E. Freitag, *A novel mutation within the central Listeria monocytogenes regulator PrfA that results in constitutive expression of virulence gene products.* J Bacteriol, 2004. **186**(18): p. 6265-76.
186. Ripio, M.T., et al., *Transcriptional activation of virulence genes in wild-type strains of Listeria monocytogenes in response to a change in the extracellular medium composition.* Res Microbiol, 1996. **147**(5): p. 371-84.
187. Mertins, S., et al., *Interference of components of the phosphoenolpyruvate phosphotransferase system with the central virulence gene regulator PrfA of Listeria monocytogenes.* J Bacteriol, 2007. **189**(2): p. 473-90.
188. Gray, M.J., N.E. Freitag, and K.J. Boor, *How the bacterial pathogen Listeria monocytogenes mediates the switch from environmental Dr. Jekyll to pathogenic Mr. Hyde.* Infect Immun, 2006. **74**(5): p. 2505-12.
189. Leimeister-Wachter, M., E. Domann, and T. Chakraborty, *The expression of virulence genes in Listeria monocytogenes is thermoregulated.* J Bacteriol, 1992. **174**(3): p. 947-52.
190. Draing, C., et al., *Cytokine induction by Gram-positive bacteria.* Immunobiology, 2008. **213**(3-4): p. 285-96.
191. Morath, S., et al., *Synthetic lipoteichoic acid from Staphylococcus aureus is a potent stimulus of cytokine release.* J Exp Med, 2002. **195**(12): p. 1635-40.

192. Rodriguez-Lazaro, D., M. Hernandez, and M. Pla, *Simultaneous quantitative detection of Listeria spp. and Listeria monocytogenes using a duplex real-time PCR-based assay.* FEMS Microbiol Lett, 2004. **233**(2): p. 257-67.
193. Dodds, A.W., *Small-scale preparation of complement components C3 and C4.* Methods Enzymol, 1993. **223**: p. 46-61.
194. Wallis, R., et al., *Molecular interactions between MASP-2, C4, and C2 and their activation fragments leading to complement activation via the lectin pathway.* J Biol Chem, 2007. **282**(11): p. 7844-51.
195. Grundling, A. and O. Schneewind, *Synthesis of glycerol phosphate lipoteichoic acid in Staphylococcus aureus.* Proc Natl Acad Sci U S A, 2007. **104**(20): p. 8478-83.
196. Fischer, W., *Physiology of lipoteichoic acids in bacteria.* Adv Microb Physiol, 1988. **29**: p. 233-302.
197. Hermann, C., *Review: variability of host-pathogen interaction.* J Endotoxin Res, 2007. **13**(4): p. 199-218.
198. Hashimoto, M., et al., *Not lipoteichoic acid but lipoproteins appear to be the dominant immunobiologically active compounds in Staphylococcus aureus.* J Immunol, 2006. **177**(5): p. 3162-9.
199. Uchikawa, K., I. Sekikawa, and I. Azuma, *Structural studies on lipoteichoic acids from four Listeria strains.* J Bacteriol, 1986. **168**(1): p. 115-22.
200. Mandin, P., et al., *VirR, a response regulator critical for Listeria monocytogenes virulence.* Mol Microbiol, 2005. **57**(5): p. 1367-80.
201. Kiriukhin, M.Y., et al., *Biosynthesis of the glycolipid anchor in lipoteichoic acid of Staphylococcus aureus RN4220: role of YpfP, the diglucosyldiacylglycerol synthase.* J Bacteriol, 2001. **183**(11): p. 3506-14.
202. Tilney, L.G. and D.A. Portnoy, *Actin filaments and the growth, movement, and spread of the intracellular bacterial parasite, Listeria monocytogenes.* J Cell Biol, 1989. **109**(4 Pt 1): p. 1597-608.
203. Dabiri, G.A., et al., *Listeria monocytogenes moves rapidly through the host-cell cytoplasm by inducing directional actin assembly.* Proc Natl Acad Sci U S A, 1990. **87**(16): p. 6068-72.
204. Lorenz, E., et al., *Relevance of mutations in the TLR4 receptor in patients with gram-negative septic shock.* Arch Intern Med, 2002. **162**(9): p. 1028-32.
205. Lorenz, E., et al., *A novel polymorphism in the toll-like receptor 2 gene and its potential association with staphylococcal infection.* Infect Immun, 2000. **68**(11): p. 6398-401.

206. Kiechl, S., et al., *Toll-like receptor 4 polymorphisms and atherogenesis.* N Engl J Med, 2002. **347**(3): p. 185-92.

# 11 List of Publications

Major parts of this thesis are published or submitted for publication in journals:

- **O. Dehus**, M. Pfitzenmaier, S. Meier, N. Fischer, G.Stübs, C. Draing, W. Schwaeble, S. Morath, T. Hartung, A. Geyer and C. Hermann: Growth temperature induces switching of structural variants of *Listeria monocytogenes* lipoteichoic acid. *Submitted*

- **O. Dehus**, G. Rogler and C. Hermann: LPS-inducible anti-inflammatory responses are not diminished in Crohn's disease patients with heterozygous Asp(299)Gly polymorphism. *Submitted*

- **O. Dehus**, S. Bunk, S. v. Aulock and C. Hermann: IL-10 release requires stronger toll-like receptor 4-triggering than TNF- a possible explanation for the selective effects of heterozygous TLR4 polymorphism Asp(299)Gly on IL-10 release. *Immunobiol 2008; 213(8):621-7.*

- **O. Dehus**, T. Hartung and C. Hermann: Endotoxin evaluation of eleven lipopolysaccharide by whole blood assay does not always correlate with Limulus Amebocyte Lysate assay. *J Endotox Res 2006; 12(3):171-80.*

12   Acknowledgements

The work presented in this thesis was carried out between February 2005 and March 2008 at the Chair of Biochemical Pharmacology, University of Konstanz, under the supervision of Prof. Dr. Dr. Thomas Hartung and PD Dr. Corinna Hermann whom I want to thank sincerely for their support.

I also thank Prof. Dr. Albrecht Wendel for the excellent working facilities provided at the Chair of Biochemical Pharmacology and for his engagement in the International Research Training Group (IRTG) 1331 (Deutsche Forschungsgemeinschaft, DFG).

I appreciate the valuable contributions from all collaborators, co-authors and technical assistants, as well as from all blood donors from the University of Konstanz and the IBD patients collective.

Significant parts of the research were funded by the DFG via the IRTG 1331 "Cell-based Characterization of Disease Mechanisms in Tissue Destruction and Repair".

Die VDM Verlagsservicegesellschaft sucht für wissenschaftliche Verlage abgeschlossene und herausragende

Dissertationen, Habilitationen, Diplomarbeiten, Master Theses, Magisterarbeiten usw.

für die kostenlose Publikation als Fachbuch.

Sie verfügen über eine Arbeit, die hohen inhaltlichen und formalen Ansprüchen genügt, und haben Interesse an einer honorarvergüteten Publikation?

Dann senden Sie bitte erste Informationen über sich und Ihre Arbeit per Email an *info@vdm-vsg.de*.

Sie erhalten kurzfristig unser Feedback!

VDM Verlagsservicegesellschaft mbH
Dudweiler Landstr. 99   Telefon +49 681 3720 174
D - 66123 Saarbrücken   Fax   +49 681 3720 1749
www.vdm-vsg.de

Die VDM Verlagsservicegesellschaft mbH vertritt

Printed by Books on Demand GmbH, Norderstedt / Germany